The Macmillan Students' Hardy
Advisory editor: James Gibson

Chosen Poems of Thomas Hardy

Other titles in the series

Far from the Madding Crowd edited by James Gibson
The Return of the Native edited by Colin Temblett-Wood
The Trumpet-Major edited by Ray Evans
The Mayor of Casterbridge edited by F. B. Pinion
The Woodlanders edited by F. B. Pinion
Tess of the d'Urbervilles edited by James Gibson
Jude the Obscure edited by Tom Wightman
Under the Greenwood Tree edited by Anna Winchcombe

Chosen Short Stories of Thomas Hardy selected and edited by James Gibson

The Macmillan Students' Hardy

Chosen Poems of Thomas Hardy

Selected and edited by James Gibson

Principal Lecturer in English at Christ Church College, Canterbury

Macmillan Education
London and Basingstoke

First published 1975
Reprinted 1976, 1978, 1980, 1982, 1983

Published by
MACMILLAN EDUCATION LTD
Houndmills Basingstoke Hampshire RG21 2XS
and London
Associated companies in Delhi, Dublin
Hong Kong, Johannesburg, Lagos, Melbourne
New York, Singapore and Tokyo

Printed in Hong Kong

Contents

Incidents and Stories

Descriptive and Animal Poems

List of Illustrations

The cover photograph is of Durdle-Door (see p. 100)

Plates between pages 80 and 81

The Life of Thomas Hardy, 1840–1928

1840 Born on 2 June at Higher Bockhampton, five kilometres from Dorchester. Began reading at an early age.

1848 Entered the new school at Lower Bockhampton. Transferred the following year to a private school at Dorchester.

1856 Became articled to John Hicks, a Dorchester architect. He described his life at this time as made up of three strands – the professional life, the scholar's life, and the rustic life.

1862 Began more advanced architectural work with a London firm. Took great interest in the cultural life of the capital, became enthusiastic about poetry, wrote many poems but failed to get any published.

1867 Returned home, assisted Hicks in church restoration, and finished the first draft of a novel by January 1868. It was never published.

1870 Visited St Juliot in Cornwall to plan the restoration of the parish church, and fell in love with Emma Gifford.

1871 *Desperate Remedies* was published. It was followed by *Under the Greenwood Tree* (1872), *A Pair of Blue Eyes* (1873), and *Far from the Madding Crowd* (1874).

1874 His growing success as a novelist encouraged Hardy and Emma Gifford to marry. During the next few years they lived in London, at Yeovil, and at Sturminster Newton, where *The Return of the Native* was written.

1878 Before this novel was published, they returned to London, where Hardy worked on *The Trumpet-Major* (1880). In 1881 they moved to Wimborne. He had written a number of short stories and was to write many more, eventually collected into four volumes.

1883 When the Hardys moved to Dorchester, two more novels had been completed. Here *The Mayor of Casterbridge* was written.

1885 They moved into Max Gate, the house on the outskirts of Dorchester which Hardy had designed for himself. Here he wrote *The Woodlanders* (1887), *Tess of the d'Urbervilles* (1891), and *Jude the Obscure* (1895).

1896 The unkind reception *Jude* received from the critics, together with a growing financial independence and perhaps the feeling that he had achieved all that he could hope to do in the novel, meant that he could now turn to his first love – poetry.

1898 *Wessex Poems*, his first book of verse, published.

1902 Began, after planning it for years, *The Dynasts*, an epic drama on the Napoleonic War. It was published in three parts from 1904 to 1908.

1910 He received the Order of Merit and the freedom of Dorchester.

1912 His wife died suddenly. Hardy, smitten with remorse for their differences and the coldness of the latter years of their marriage, wrote many poems (over a long period) about his recollections of her.

1914 Married Florence Dugdale. The outbreak of the First World War made him regret having ended *The Dynasts* optimistically for the future of mankind.

1928 He died on January 11. His last volume of verse, his eighth, appeared posthumously. Although his first books of verse had been indifferently received, he was by now regarded not only as a great novelist but as a great poet.

Hardy was born when Queen Victoria, who became Queen in 1837, was twenty-one. His life covers most of her long reign, that of Edward VII, and most of the reign of George V. He lived through nearly a century of increasingly rapid change. When he was born the railway had not yet reached Dorset, and public hangings were common. His grandmother talked to him about her memories of the French Revolution and the Napoleonic War, and he grew up in a small, closely-knit rural community which had seen little change for centuries. Hardy himself was to see that community radically altered by the impact of new ideas and inventions. Modern methods of travel and communication destroyed the traditions and the isolation and unity of the old village life, and brought about big changes in outlook and behaviour.

Scientific discoveries had led to a questioning of religious beliefs seldom doubted before. The simple faith of Hardy's parents, who took him as a boy to Stinsford church, was not possible for Hardy, who, like other distinguished Victorian writers, found himself torn between a desire to believe and an intellectual inability to do so. His wide reading and great intelligence made him aware of these changes; his honesty would not allow him to dodge their implications.

His loss of faith was one of the reasons for the sadness which permeates a great deal of his writing. Yet he consistently believed that the maintenance of Christian virtues was the only hope for mankind in the modern scientific age. He had a sensitive awareness of history and of the transience of human life, and much of his best poetry springs from this. It deals with events as remote as the Roman occupation of Britain and as recent for him as the sinking of the *Titanic* and the holocaust of the First World War. His novels were all written in the nineteenth century and deal with many aspects of the immediate past of that part of England, centred in Dorset, which for historical reasons he chose to call 'Wessex'. To appreciate Hardy's work fully it is necessary to share his sense of history, his awareness of man's place in a universe which seems indifferent to him, and his conviction that what he liked to call 'loving-kindness' must govern our behaviour if there is to be any hope for the human race.

Introduction

When in 1898 Thomas Hardy published his first book of verse, *Wessex Poems*, he was already famous as a novelist. It was the reception which his novel, *Jude the Obscure*, had received in 1896 which had finally changed him from novelist to poet. In *Jude* he had tried to deal in a serious way with the agony of a man torn between two women – one a slut and the other an intellectual. Hardy's frank treatment of the subject was condemned by some of the most powerful Victorian critics and he was viciously and unfairly attacked as an immoral writer and an opponent of marriage. Anonymous and libellous letters were sent to him at his home at Max Gate, and he even received a letter containing a packet of ashes which were said to be those of his wicked book. Always a sensitive man, he was very hurt by these scurrilous attacks and he turned to poetry in the hope that, as he wrote in his journal:

> I can express more fully in verse ideas and emotions which run counter to the inert crystallized opinion – hard as a rock – which the vast body of men have vested interests in supporting . . . If Galileo had said in verse that the world moved, the Inquisition might have left him alone.

HARDY'S LIFE AS A POET

He had always loved poetry and he believed, as he once wrote, that 'in verse is concentrated the essence of all imaginative and emotional literature'. His first poem known to have survived, 'Domicilium' (p. 29), is a description of his cottage home at Higher Bockhampton and was written when he was about eighteen. His ambition was to be a writer and while in London in the 1860s practising as an architect he had often written poems which he tells us he had failed to get published. Always a careful and methodical man, he kept some of these poems by him and many years later they were to appear, sometimes in a revised form, in the books published as his fame as a poet grew. It is not easy to arrange Hardy's poems in a strict chronological order because of his habit of unearthing poems written many years earlier and publishing them sometimes, but not always, with the date on which they were originally written.

There can be few poets who have published their first book of verse when they were nearly sixty, but for Hardy this was just the beginning. When he died in 1928 he had published a further six books of verse, and a last book, *Winter Words*, was published shortly after his death. He had hoped that he might live to be the only English poet ever to publish a book of poems on his ninetieth birthday. Most of these eight books of verse contain at least one hundred poems, and in his *Collected Poems* published in 1930 there were nearly a thousand poems – a remarkable testimony to the creativity of this remarkable man. Quantity is not a virtue, but it is of interest that as distinguished a poet and critic as Philip Larkin has said of the *Collected Poems*:

> I love the great Collected Hardy which runs for something like 800 pages. One can read him for years and years and still be surprised, and I think that's a marvellous thing to find in any poet.

No selection of Hardy's verse can do justice to this creative miracle, and it is to the *Collected Poems*, which have been continuously in print for nearly fifty years, that the real lover of the poetry must eventually go.

Success as a poet did not come that easily. Hardy was so uncertain of the welcome that *Wessex Poems* would receive that he told his publishers that he would take on his shoulders the financial risk of producing the book. His publishers, to their credit, did not accept his offer, and the book had a respectful but not rapturous reception. Some reviewers thought that he would have been wiser to stick to novel-writing, but this did not deter him from his intention to be a poet, and, after the publication of another book, *Poems of the Past and the Present*, in 1902, he worked for several years on his great poetry epic, *The Dynasts*, which was published between 1904 and 1908. This is a very long poem in three books and nineteen acts about the Napoleonic wars. Its success was considerable, and he soon found himself accepted as both a distinguished novelist and a distinguished poet. Newspapers and journals began to compete for his poetry, and his subsequent books of verse were printed in ever-larger editions. But fame did not affect Hardy, and it was typical of his modesty that when he sent a poem to a journal he often included a stamped and addressed envelope in case the editor should wish to reject it.

When Hardy died in 1928 the obituary notices spoke of him not only as a great novelist but also as a great poet. There were even those who thought that his achievement as a poet was the greater and that it was as a poet that he would be finally remembered. Now, almost fifty years later, we can see that part of his uniqueness as a writer is that he was indisputably great in both poetry and prose.

THE GREATNESS OF HARDY'S POETRY

What do we mean by 'greatness' in poetry and what possible relevance can Hardy's poetry have for us today? In an important sense these two questions are connected because it is a feature of all truly great writing that it transcends its own time and remains relevant for future generations. This doesn't mean that the poet sets out to write about vague immortal subjects in a remote and artificial language. He doesn't. Nobody could have his feet more firmly on the ground than Hardy. What it means is that whatever subject he is dealing with he is able through the strength of his own feelings and his powers of expression to make us – far away from him in time and space – share his feelings. This is what we mean when we talk about the 'universality' of a great writer's work. Shakespeare and Wordsworth and Hardy may have lived in worlds very different from ours today, but what we have in common with them is far more important than what is different. The internal is what matters finally – not the external. Whether we are old or young, black or white, rich or poor, live in the sixteenth century or the twentieth, we all have certain feelings that we share. Most of us fall in love, have (most of the time!) love for our parents and children, are capable of feeling sympathy, of experiencing happiness and sadness, and, in fact, feeling in many vital ways much as the great writers of the past convince us they felt. The future for mankind depends upon our realising how much we all have in common, and great writers like Hardy saw this and help us to see it. When he talks about religion and poetry touching each other, as he does in the introduction to one of his books of verse, this is what he means. In the middle of the bloodiest war of all time Hardy was able to write:

Yonder a maid and her wight
  Come whispering by:
War's annals will cloud into night
  Ere their story die.

We feel with Hardy the inescapable truth of this. We are separated by war but joined by our ability to feel love. War is destructive but love is creative, and in spite of wars men and women will continue to fall in love.

HARDY'S UNIVERSALITY

If you read the poem on p. 81 from which that stanza comes, you will see how simply and yet powerfully, in a bare twelve lines, Hardy has said something which is of a universal nature, as true now as when it was written, as true in Africa as in Europe. The tractor has taken the place of the horse, but our deeper needs still require us to be creative, to produce food, to fall in love, to have children. Life seems to become ever more complex, but Hardy's poem by its very simplicity says to us that it is the simple, lasting things which really matter. This is true relevance as opposed to the false relevance which assumes that anything which is about life today, particularly if it is about a fish-and-chip shop or a kitchen sink or about politics, *must* be relevant, and that anything from the past *must* be irrelevant. A great contemporary of Hardy, the Irish poet, W. B. Yeats, in a poem which has much in common with Hardy's poem, puts it very well:

How can I, that girl standing there,
My attention fix
On Roman or on Russian
Or on Spanish politics?

Yet here's a travelled man that knows
What he talks about,
And there's a politician
That has read and thought,
And maybe what they say is true
Of war and war's alarms,
But O that I were young again
And held her in my arms!

You don't have to be old to sympathise with how Yeats felt.

Let us look at another poem by Hardy in which we see the same universality at work. It is so typical of Hardy that it is worth quoting in full and looking at in some detail:

While rain, with eve in partnership,
Descended darkly, drip, drip, drip,
Beyond the last lone lamp I passed
  Walking slowly, whispering sadly,
  Two linked loiterers, wan, downcast:
Some heavy thought constrained each face
And blinded them to time and place.

The pair seemed lovers, yet absorbed
In mental scenes no longer orbed
By love's young rays. Each countenance
  As it slowly, as it sadly
  Caught the lamplight's yellow glance,
Held in suspense a misery
At things which had been or might be.

When I retrod that watery way
Some hours beyond the droop of day,
Still I found pacing there the twain
  Just as slowly, just as sadly,
  Heedless of the night and rain.
One could but wonder who they were
And what wild woe detained them there.

Though thirty years of blur and blot
Have slid since I beheld that spot,
And saw in curious converse there
  Moving slowly, moving sadly
  That mysterious tragic pair,

Its olden look may linger on
All but the couple; they have gone.

Whither? Who knows, indeed. . . . And yet
To me, when nights are weird and wet,
Without those comrades there at tryst
  Creeping slowly, creeping sadly,
  That lone lane does not exist.
There they seem brooding on their pain,
And will, while such a lane remain.

Again it is about love and the passing of time – Hardy's favourite subjects – but this time the love is unhappy. One of the remarkable features of this and of many of Hardy's poems is how slight and ordinary the incident was that eventually led to the writing of the poem. He goes for a walk in the evening and it is raining. Under the lamplight he sees a pair of lovers obviously in some distress. On his way back they are still there 'heedless of the night and rain'. It takes a great poet to make a memorable poem out of an incident as slight as that, an incident such as many of us will have witnessed and not given another thought – but then we are not poets.

How does Hardy make such an incident into a poem? Again we go first to its universal quality. It is not saying anything as profoundly simple as the poem about love and war, but its subject, the sorrows of love, is something most of us will have experienced in one form or another in our lives, and the poem's universality comes from that. We don't know who this unhappy pair are, they and their particular grief have been dead these many years, but by his art as a poet Hardy has immortalised their sorrow and we can share it and feel pity for them (and for ourselves?) almost a hundred years later. But we wouldn't feel like that unless we sensed that Hardy's poetic experience was genuine, and it is another part of his greatness that what he writes in his poems always has the stamp of truth. It is difficult to doubt that this experience really happened. It is, as we have said, the kind of incident that does happen and Hardy provides us with enough circumstantial evidence to make it seem genuine. We are told that it happened in Tooting, where Hardy was living in 1880, that it was raining, that the poet was going for his evening walk. But a great deal is deliberately left uncertain so that this unhappy pair can enter our imagination and appeal to our sympathy. We are not told why they are unhappy (Hardy doesn't know and it is much better left to our imagination) and there is a shadowiness about them which seems to make them into a symbol of all unhappy couples everywhere. This is made more powerful by the title 'Beyond the Last Lamp', with its suggestion that the couple are not only literally but also metaphorically going into the dark. Hardy's first title for the poem was 'In the Suburbs'. You will see how much he has gained imaginatively by the change.

The poem is given even greater meaning by the discovery in stanza 4 that the incident, described as if it has just happened, took place over thirty years before. By this kind of double vision by which he sees the past as if it were the present, and then distances it as if he were taking a telescope away from his eye (he uses the same technique in 'Beeny Cliff' on p. 74) Hardy puts all our sufferings against a massive background of passing time which he sees as the enemy of us all, as it is. You will notice how often his theme is what has been called the sadness of things, the sadness of looking back at times that have passed away and of being aware that even our happiest moments cannot last: we cannot stop time. Time, then, is important in Hardy's poetry, and so is place, and his final thought in this poem is that it is those tragic figures in that street which gave the place its meaning. If you return to a school you have left you will realise what Hardy means. It is the same place and yet different because the people you knew there and who gave the place a meaning for you are no longer there. For Hardy people come first – and this is another sign of his greatness. He once wrote that 'the beauty of association is entirely superior to the beauty of aspect, and a beloved relative's old battered tankard to the finest Greek vase'.

The incident, then, is slight and in no way unusual, and very little happens in the five stanzas, yet the poem is packed with feeling. There is the unknown sorrow of the lovers, the sadness of passing time, and, most important of all, Hardy's great pity for others who are in pain, a pity which, if the poem works for you, you will feel too. Hardy was very sensitive to suffering, whether it was of an old workman in pain as a result of his labours, unhappy lovers, those who mourn for the dead, the probable future sufferings of an unborn pauper child, or a robin dying from cold and hunger. A sensitive reading of Hardy's poems will reveal the presence behind them of a man of unswerving honesty of vision, of passionate sincerity, and of powerful feelings.

HARDY'S CRAFTSMANSHIP

But feelings alone do not make a great poem. There must also be a distinctive control of the techniques of writing verse. The good poet must not only be able to feel, he must be able to express what he feels in memorable language and a unique style. It has been said by critics who are unsympathetic to Hardy's verse (but insist on writing about it) that he is inept in his use of language and that his rhythms show that he had a defective ear. 'Beyond the Last Lamp' and many other poems in this book will give the lie to that. What may at first hearing seem awkward will nearly always prove on further acquaintance to be remarkably apt and to be typical of Hardy's style in his determination to find exactly the right words to describe his experience and feelings. For example, 'linked loiterers' is unusual and may strike us at first as awkward. One suspects that Hardy may have been attracted to this unusual description by his love of alliteration. But the very unusualness of this combination of words makes the reader think about it, and then it becomes clear that a great deal of meaning is packed into those two words. 'Linked' has physical associations – the lovers have their arms around each other – and also has the metaphorical meaning of two people joined together by their love for each other, possibly even joined together in their troubles as if by the links of a chain. And 'loiterers' has the sense of their aimless movement backward and forward under the lamplight as they talk over the causes of their grief. You may have noticed that people in distress do find it difficult to stay still. It is always worth looking very closely at Hardy's use of language. His long period of novel writing had made him a craftsman in words. In looking at this poem we should think deeply about the meanings of such words as the 'partnership' of rain and evening, the 'lone' lamp, love's young 'rays' (think of the association of rays with a lamp), and 'blur and blot', to mention just some of the really interesting words. You will soon discover that Hardy is a far more sensitive user of language than are some of the literary critics who call him awkward. You will find, too, as you read through his poems that inspired descriptions like 'time-torn man' (p. 53), 'mothy and warm' (p. 87), 'morning harden upon the wall' (p. 67), 'the hope-hour' (p. 52), and 'bulks old Beeny to the sky' (p. 74) stay in the memory because of their rich suggestiveness and provide yet more evidence of Hardy's claim to be a major poet.

His rhythms, too, are anything but awkward. He said himself that there was a natural music in the sincere language of the emotions and we hear this music throughout his verse. 'Beyond the Last Lamp' is full of music if read aloud – as all Hardy's poetry should be – and though its rhythm is basically regular, there are shrewdly placed variations in the beat which remove any possibility of monotony. Note, for example, how after the first three lines of the first stanza in which the first strong accent comes on the second syllable, the first strong accent in the fourth line is on the first syllable. The refrain '. . . slowly . . . sadly' is cleverly used to emphasise the mood of wistfulness, and yet Hardy keeps changing the basic structure of the line (see how strongly it runs on into the next line in stanza 2) so that here again it does not become monotonous. At the same time the words 'slowly' and 'sadly' act as rhymes joining all the stanzas together and emphasising the unity of the poem and of the experience, even though thirty years separate the two parts of the poem. This effect of unity is strengthened and our pleasure in the sound is increased by the way in which the sound of the word 'rain' in the first line is picked up in line 6 in 'constrained', and then with 'twain' and 'rain' becomes a rhyme in stanza 3, to be repeated yet again in 'detained' in the last line of that stanza. Finally, like a musical chord, it is heard again very dominantly at the very end of the poem, where it appears four times in the last three lines, twice in the word 'lane', and then in the culminating rhyme 'pain' and 'remain' Hardy weaves a pattern of sound which makes us aware of the artistic beauty of the poem he has constructed at the same time as we feel the sadness of what is being described.

If you really look closely at a Hardy poem you will find that he is a very clever and careful craftsman. Just notice how many different stanza shapes he uses in this selection alone, and each shape is a piece of craftsmanship carefully designed for a particular purpose. He came from a long line of Dorset craftsmen – his father and grandfather were builders and stone-masons – who took pride in their work, and Hardy carried on in his writing their traditions of doing a job well. You will have to look *closely* because his is an art which conceals art. Very many of the characteristics of Hardy's poetry which we have seen in 'Beyond the Last Lamp' are to be found in other poems in this book, and it will not be long before you will be able to recognise one of his poems by its highly individualistic flavour. A great writer's personality and style are always distinguishable from those of other writers.

HARDY'S RANGE

There are other features of his poetry which deserve mention. There is the sheer range of subject matter. It has been said of Hardy that he is great only in respect of about a dozen poems written after the death of his first wife. These poems do, indeed, represent the summit of his poetry, but it would be stupid to regard everything else as second-class. Is it likely that a man who wrote twelve great poems should write nearly another thousand that were definitely not great? That it was not just the death of his wife which could inspire him to write great poetry is clear from any study of his verse. Where is there to be found a greater poem about the waste and sadness of war than 'Drummer Hodge' (p. 56) written, incidentally, years before Sassoon and Owen wrote their poems about the pity and horror of war? Is there a finer occasional poem – that is one written about a public event – than 'The Convergence of the Twain' (p. 63), or a greater poem about man's need for something to believe in than 'The Oxen'

(p. 33), or a more moving poem about the death of a pet than 'Last Words to a Dumb Friend' (p. 144)? And where are there to be found better narrative poems than those you will find in the narrative section of this book? Hardy loved strange and exciting stories and he had been brought up on the old ballads with their drama and sudden death, their strong rhythms and wistful refrains. An excellent example is 'A Trampwoman's Tragedy' (p. 111). Hardy himself thought that this was one of his most successful poems. There are, too, the straightforward descriptive poems which he wrote so well about the country and the country life. He not only loved the country but he observed it most sensitively, and acute powers of observation are another gift of the great poet. Having been brought up in the country and knowing it in all its moods, he didn't sentimentalise it as so many poets have done. He sees it in good weather and bad (p. 127), he knows that the 'merry month of May' is not always merry (p. 130) and that the leaves tremble in October (p. 133). His country people are real people (p. 135) who, no matter what the weather, must go to market for provisions and count their sheep in the 'sour spring wind' (p. 130). Here in all these poems on all these different aspects of life is 'God's plenty' and it is not surprising that a critic described this great span of poetry as an 'imaginative record or commentary of life'. That describes it exactly. There is nothing narrow about Hardy. His canvas is the whole of life and there is no subject so small that he cannot make poetry out of it.

PERSONAL QUALITIES

Yet though his range is so wide, most of Hardy's poetry is very personal. Some of it is obviously about himself, about his parents, his childhood, his marriage, the death of his wife, and his growing old. And even when he is writing about subjects which do not seem directly related to him personally, the reader senses that he is very much involved. This very personal nature of his writing is not liked by some critics, possibly because with it goes a lack of the intellectual abstractions, scholarly allusions, and critical difficulties which in the poetry of someone like T. S.

Eliot provide so many opportunities for the writing of books of endless commentary and explanation. But for many of us the comparative clarity and directness of Hardy's approach, the simple honesty and sincerity of outlook, his gentleness and humility, and the immediate feeling we have of his deep concern and sympathy for others will more than atone for the lack of more academic qualities. As we read his poetry we feel that we are listening to another human being very like ourselves in spite of his great gifts. He is so much one of the people, intensely interested in people, and we, as people ourselves, feel an interest in him. He himself in one of his notebooks quoted approvingly this belief of another Victorian writer:

> The ultimate aim of the writer should be to touch our hearts by showing his own, and not to exhibit his learning or his fine taste.

When all is said, it remains for you to judge to what extent Hardy is successful in touching your heart.

Editorial Note

Editors have not found it easy to arrange Hardy's poetry in a meaningful and orderly way. Hardy's own division into 'Poems of Place and Incident' and 'Poems of Memory and Reflection' is unsatisfactory, because most of his poems of place and incident are also poems of memory and reflection. For this edition I have arranged my selection in a completely new way resulting from my belief that, as he is one of the most autobiographical of poets, it should be possible to arrange his poems in a broadly chronological way based upon the period with which the poem is concerned. The result is still not completely satisfactory but does have the great advantage that it tells in an orderly way a great deal about the main incidents in Hardy's life, and the poetry is seen to emerge from the life. Two smaller groups of poems would not fit into this arrangement. The large opening section which is entitled 'Poems Mainly Autobiographical' is followed, therefore, by two smaller sections; 'Incidents and Stories' (p. 89), and 'Descriptive and Animal Poems' (p. 125).

No editor of Hardy's poetry could fail to be in debt to Professor J. O. Bailey. His *The Poetry of Thomas Hardy* (The University of North Carolina Press) is a valuable work of reference. The most useful complete critical work on the poetry is Kenneth Marsden's *The Poems of Thomas Hardy* (Athlone Press). Helpful chapters will also be found in Douglas Brown's *Thomas Hardy* (Longman), R. P. Blackmur's *Language as Gesture*, Samuel Hynes's *The Pattern of Hardy's Poetry* (The University of North Carolina Press), and Trevor Johnson's *Thomas Hardy* (Evans). References in the Notes to the *Life* are to *The Life of Thomas Hardy* by F. E. Hardy (Macmillan).

Poems Mainly Autobiographical

The Field of Waterloo
(From *The Dynasts*)

YEA, the coneys are scared by the thud of hoofs,
And their white scuts flash at their vanishing heels,
And swallows abandon the hamlet-roofs.

The mole's tunnelled chambers are crushed by wheels,
The lark's eggs scattered, their owners fled; [5]
And the hedgehog's household the sapper unseals.

The snail draws in at the terrible tread,
But in vain; he is crushed by the felloe-rim;
The worm asks what can be overhead,

And wriggles deep from a scene so grim, [10]
And guesses him safe; for he does not know
What a foul red rain will be soaking him!

Beaten about by the heel and toe
Are butterflies, sick of the day's long rheum,
To die of a worse than the weather-foe. [15]

Trodden and bruised to a miry tomb
Are ears that have greened but will never be gold,
And flowers in the bud that will never bloom.

As a boy Hardy was fascinated by the Napoleonic War which ended only twenty-five years before his birth. His grandmother talked to him about her memories of the threat of invasion by Napoleon: his father had among his books a history of the War. *The Dynasts*, published between 1904 and 1908, is Hardy's great long poem about that period of history. The Battle of Waterloo was fought in 1815 and after bloody fighting Napoleon was finally defeated.

[1] **coneys** rabbits.
[2] **scuts** tails.
[6] **sapper** engineer.
[8] **felloe-rim** the rim of the wheel of the gun-carriage.
[14] **rheum** wetness.

One We Knew

(M.H. 1772–1857)

SHE told how they used to form for the country dances –
  'The Triumph', 'The New-rigged Ship' –
To the light of the guttering wax in the panelled manses,
  And in cots to the blink of a dip.

She spoke of the wild 'poussetting' and 'allemanding' [5]
  On carpet, on oak, and on sod;
And the two long rows of ladies and gentlemen standing,
  And the figures the couples trod.

She showed us the spot where the maypole was yearly planted,
  And where the bandsmen stood [10]
While breeched and kerchiefed partners whirled, and panted
  To choose each other for good.

She told of that far-back day when they learnt astounded
  Of the death of the King of France:
Of the Terror; and then of Bonaparte's unbounded [15]
  Ambition and arrogance.

Of how his threats woke warlike preparations
  Along the southern strand,
And how each night brought tremors and trepidations
  Lest morning should see him land. [20]

She said she had often heard the gibbet creaking
  As it swayed in the lightning flash,
Had caught from the neighbouring town a small child's shrieking
  At the cart-tail under the lash. . . .

With cap-framed face and long gaze into the embers – [25]
  We seated around her knees –
She would dwell on such dead themes, not as one who remembers,
  But rather as one who sees.

She seemed one left behind of a band gone distant
  So far that no tongue could hail: [30]
Past things retold were to her as things existent,
  Things present but as a tale.

20 May, 1902

M.H. was Mary Hardy, the poet's grandmother who died in the Hardys' cottage at Higher Bockhampton in January 1857. Hardy inherited her love of music, her interest in the Napoleonic War, and her interest in the past.

[3] **manses** mansions.
[4] **dip** candle.
[5] **'poussetting' . . . 'allemanding'** names of dances.
[11] **kerchiefed** A kerchief was a piece of cloth worn as a kind of hat by women.
[15] **the Terror** the French Revolution.
Bonaparte i.e. Napoleon.
[21] **gibbet** gallows.

Domicilium

It faces west, and round the back and sides
High beeches, bending, hang a veil of boughs,
And sweep against the roof. Wild honeysucks
Climb on the walls, and seem to sprout a wish
(If we may fancy wish of trees and plants) [5]
To overtop the apple-trees hard by.

Red roses, lilacs, variegated box
Are there in plenty, and such hardy flowers
As flourish best untrained. Adjoining these
Are herbs and esculents; and farther still [10]
A field; then cottages with trees, and last
The distant hills and sky.

Behind, the scene is wilder. Heath and furze
Are everything that seems to grow and thrive
Upon the uneven ground. A stunted thorn [15]

Stands here and there, indeed; and from a pit
An oak uprises, springing from a seed
Dropped by some bird a hundred years ago.

In days bygone –
Long gone – my father's mother, who is now [20]
Blest with the blest, would take me out to walk.
At such a time I once inquired of her
How looked the spot when first she settled here.
The answer I remember. 'Fifty years
Have passed since then, my child, and change has marked [25]
The face of all things. Yonder garden-plots
And orchards were uncultivated slopes
O'ergrown with bramble bushes, furze and thorn:
That road a narrow path shut in by ferns,
Which, almost trees, obscured the passer-by. [30]

'Our house stood quite alone, and those tall firs
And beeches were not planted. Snakes and efts
Swarmed in the summer days, and nightly bats
Would fly about our bedrooms. Heathcroppers
Lived on the hills, and were our only friends; [35]
So wild it was when first we settled here.'

Hardy wrote this poem when he was in his late teens. It describes the cottage at Higher Bockhampton in which his family had lived since 1801, in which he had been born in 1840, and where he lived for most of the first thirty years of his life. This is the earliest piece of Hardy's verse known to have survived. It shows the influence of the poet, Wordsworth. His 'father's mother' was Mary Hardy about whom he writes in 'One We Knew'.

[10] **esculents** vegetables.
[32] **efts** small lizards.
[34] **Heathcroppers** heath ponies.

A Church Romance
(*Mellstock:* about 1835)

SHE turned in the high pew, until her sight
Swept the west gallery, and caught its row
Of music-men with viol, book, and bow
Against the sinking sad tower-window light.

She turned again; and in her pride's despite [5]
One strenuous viol's inspirer seemed to throw
A message from his string to her below,
Which said: 'I claim thee as my own forthright!'

Thus their hearts' bond began, in due time signed.
And long years thence, when Age had scared Romance, [10]
At some old attitude of his or glance
That gallery-scene would break upon her mind,
With him as minstrel, ardent, young, and trim,
Bowing 'New Sabbath' or 'Mount Ephraim'.

'Mellstock' is Hardy's name for Stinsford, the parish in which Higher Bockhampton lies. Hardy's father and grandfather played musical instruments in Stinsford Church choir. Music played a large part in Hardy's youth and greatly influenced his poetry. Note how this sonnet is divided by time into two – the romance (the first eight lines) and looking back on the romance years later (the last six lines). This is a favourite Hardy structure.

[1] **She** Hardy's mother.
[3] **viol** a stringed musical instrument played with a bow.
[8] **forthright** straightaway.
[14] **'New Sabbath ...'Mount Ephraim'** very popular hymn tunes of the time.

The Roman Road

THE Roman Road runs straight and bare
As the pale parting-line in hair
Across the heath. And thoughtful men
Contrast its days of Now and Then,
And delve, and measure, and compare; [5]

Visioning on the vacant air
Helmed legionaries, who proudly rear
The Eagle, as they pace again
The Roman Road.

But no tall brass-helmed legionnaire [10]
Haunts it for me. Uprises there
A mother's form upon my ken,
Guiding my infant steps, as when
We walked that ancient thoroughfare,
The Roman Road. [15]

Dorset is full of history. A Roman road from Dorchester passes close to the cottage at Bockhampton, and then crosses the heath which stretches away to the east behind it. Hardy always saw people as more important than the landscape in which they appeared.

[8] **Eagle** The Roman legions carried as their ensign the figure of an eagle.
[12] **ken** sight.

Childhood among the Ferns

I SAT one sprinkling day upon the lea,
Where tall-stemmed ferns spread out luxuriantly,
And nothing but those tall ferns sheltered me.

The rain gained strength, and damped each lopping frond,
Ran down their stalks beside me and beyond, [5]
And shaped slow-creeping rivulets as I conned,

With pride, my spray-roofed house. And though anon
Some drops pierced its green rafters, I sat on,
Making pretence I was not rained upon.

The sun then burst, and brought forth a sweet breath [10]
From the limp ferns as they dried underneath:
I said: 'I could live on here thus till death';

And queried in the green rays as I sate:
'Why should I have to grow to man's estate,
And this afar-noised World perambulate?' [15]

Hardy's love of the country and wish that time could stand still dominate this poem. His sense of the 'on-going' of time was the source of much of his sadness, as it is for many of us, but it is unusual for this sense to develop as early as it did in Hardy. In his *Life* (p. 15) he tells us how when he was less than eight years old, 'He was lying on his back in the sun' and 'he came to the conclusion that he did not wish to grow up.' It seems likely that this poem based upon that childhood memory was not written until Hardy was in his eighties. His remarkable retentive memory seemed capable of bringing the distant past into the present as if the intervening years had not existed.

[1] **lea** meadow.
[4] **frond** the leaf of a fern.

The Oxen

CHRISTMAS EVE, and twelve of the clock.
  'Now they are all on their knees,'
An elder said as we sat in a flock
  By the embers in hearthside ease.

We pictured the meek mild creatures where [5]
  They dwelt in their strawy pen,
Nor did it occur to one of us there
  To doubt they were kneeling then.

So fair a fancy few would weave
  In these years! Yet, I feel, [10]
If someone said on Christmas Eve,
  'Come; see the oxen kneel

'In the lonely barton by yonder coomb
  Our childhood used to know,'
I should go with him in the gloom, [15]
  Hoping it might be so.

1915

As a child Hardy went regularly to church but he lost his faith as he grew older. Throughout his life there was a conflict between his wish to believe and his inability to do so. Note how well this conflict is shown by the tension in the word 'Hoping'.

[13] **barton** farmyard.
coomb hollow or wooded valley.

The Colour

(The following lines are partly original, partly remembered from a Wessex folk-rhyme)

'WHAT shall I bring you?
Please will white do
Best for your wearing
  The long day through?'
'– White is for weddings, [5]
Weddings, weddings,
White is for weddings,
  And that won't do.'

'What shall I bring you?
Please will red do [10]
Best for your wearing
  The long day through?'

'– Red is for soldiers,
Soldiers, soldiers,
Red is for soldiers, [15]
  And that won't do.'

'What shall I bring you?
Please will blue do
Best for your wearing
  The long day through?' [20]
'– Blue is for sailors,
Sailors, sailors,
Blue is for sailors,
  And that won't do.'

'What shall I bring you? [25]
Please will green do
Best for your wearing
  The long day through?'
'– Green is for mayings,
Mayings, mayings, [30]
Green is for mayings,
  And that won't do.'

'What shall I bring you
Then? Will black do
Best for your wearing [35]
  The long day through?'
'– Black is for mourning,
Mourning, mourning,
Black is for mourning,
  And black will do.' [40]

Ballads and folk-songs and folk-rhymes played a large part in Hardy's growing up and strongly influenced his own poetry. Hardy, who probably remembered this rhyme from his own childhood would have been delighted to know that more than a hundred years later children in a school in Kent were heard in the playground chanting a counting-out rhyme which ended, 'Red is for soldiers, soldiers, soldiers, Red is for soldiers and out go you.'

Wessex Hardy's name for certain of the southern and western counties of England.
[29] **mayings** celebrating May Day, the first day of May.

Afternoon Service at Mellstock
(About 1850)

ON afternoons of drowsy calm
We stood in the panelled pew,
Singing one-voiced a Tate-and-Brady psalm
To the tune of 'Cambridge New'.

We watched the elms, we watched the rooks, [5]
The clouds upon the breeze,
Between the whiles of glancing at our books,
And swaying like the trees.

So mindless were those outpourings! –
Though I am not aware [10]
That I have gained by subtle thought on things
Since we stood psalming there.

Yet another memory of his childhood which Hardy uses to make the point that the thinker is not necessarily happier than those who just enjoy life as it comes without thinking too much about it.

Mellstock Hardy's name for Stinsford, the parish in which he lived.
[3] **Tate-and-Brady** Nahum Tate (1652–1715) and Nicholas Brady (1659–1726) were minor poets who produced a famous metrical version of the Psalms.

A Wet Night

I PACE along, the rain-shafts riddling me,
Mile after mile out by the moorland way,
And up the hill, and through the ewe-leaze gray
Into the lane, and round the corner tree;

Where, as my clothing clams me, mire-bestarred, [5]
And the enfeebled light dies out of day,
Leaving the liquid shades to reign, I say,
'This is a hardship to be calendared!'

Yet sires of mine now perished and forgot,
When worse beset, ere roads were shapen here, [10]
And night and storm were foes indeed to fear,
Times numberless have trudged across this spot
In sturdy muteness on their strenuous lot,
And taking all such toils as trifles mere.

From 1849 until 1862 Hardy, first as a schoolboy and then as an architectural apprentice, walked daily to and from Dorchester, nearly three miles from Bockhampton. Again we have the feeling for the past, and the use of a sonnet which divides into two sections, but this time it is the first section which deals with the present, the second with the past.

[3] **ewe-leaze** sheep pasture where, at lambing time, ewes are separated from other sheep.

The Self-Unseeing

HERE is the ancient floor,
Footworn and hollowed and thin,
Here was the former door
Where the dead feet walked in.

She sat here in her chair, [5]
Smiling into the fire;
He who played stood there,
Bowing it higher and higher.

Childlike, I danced in a dream;
Blessings emblazoned that day; [10]
Everything glowed with a gleam;
Yet we were looking away!

Hardy published this poem in 1901. It describes a return visit to his childhood home at Higher Bockhampton where he has a nostalgic vision of the past. The 'dead feet' are those of his father who died in 1892 and it is probably his father playing the violin while Hardy dances and his mother sits in her chair. Looking back now to his childhood it seems such a happy time and yet somehow he did not realise how happy that present, now the past, was.

Old Furniture

I KNOW not how it may be with others
  Who sit amid relics of householdry
That date from the days of their mothers' mothers,
  But well I know how it is with me
    Continually. [5]

I see the hands of the generations
  That owned each shiny familiar thing
In play on its knobs and indentations,
  And with its ancient fashioning
    Still dallying: [10]

Hands behind hands, growing paler and paler,
  As in a mirror a candle-flame
Shows images of itself, each frailer
  As it recedes, though the eye may frame
    Its shape the same. [15]

On the clock's dull dial a foggy finger,
  Moving to set the minutes right
With tentative touches that lift and linger
  In the wont of a moth on a summer night,
    Creeps to my sight. [20]

On this old viol, too, fingers are dancing –
  My father's – just over the strings by the nut,
The tip of a bow receding, advancing
  In airy quivers, as if it would cut
    The plaintive gut. [25]

And I see a face by that box for tinder,
  Glowing forth in fits from the dark,
And fading again, as the linten cinder
  Kindles to red at the flinty spark,
    Or goes out stark. [30]

Well, well. It is best to be up and doing,
  The world has no use for one today
Who eyes things thus – no aim pursuing!
  He should not continue in this stay,
    But sink away. [35]

It is not the furniture itself that is important to Hardy but its association with people, here his family. He once wrote that 'a beloved relative's old battered tankard is entirely superior to the finest Greek vase.'

[19] **In the wont of** like.
[22] **nut** the mechanism for tightening or slackening a violin bow.
[26] **box for tinder** the box containing the flint and steel from which the spark was struck.
[28] **linten** combustible linen material.

To Lizbie Browne

DEAR Lizbie Browne,
Where are you now?
In sun, in rain? –
Or is your brow
Past joy, past pain, [5]
Dear Lizbie Browne?

Sweet Lizbie Browne,
How you could smile,
How you could sing! –
How archly wile [10]
In glance-giving,
Sweet Lizbie Browne!

And, Lizbie Browne,
Who else had hair
Bay-red as yours, [15]
Or flesh so fair
Bred out of doors,
Sweet Lizbie Browne?

When, Lizbie Browne,
You had just begun [20]
To be endeared
By stealth to one,
You disappeared
My Lizbie Browne!

Ay, Lizbie Browne, [25]
So swift your life,
And mine so slow,
You were a wife
Ere I could show
Love, Lizbie Browne. [30]

Still, Lizbie Browne,
You won, they said,
The best of men
When you were wed. . . .
Where went you then, [35]
O Lizbie Browne?

Dear Lizbie Browne,
I should have thought,
'Girls ripen fast,'
And coaxed and caught [40]
You ere you passed,
Dear Lizbie Browne!

But, Lizbie Browne,
I let you slip;
Shaped not a sign; [45]
Touched never your lip
With lip of mine,
Lost Lizbie Browne!

So, Lizbie Browne,
When on a day [50]
Men speak of me

As not, you'll say,
'And who was he?' –
Yes, Lizbie Browne!

In Hardy's *Life* (p. 25) we are told, 'there was another young girl, a gamekeeper's pretty daughter, who won Hardy's boyish admiration because of her beautiful bay-red hair. But she despised him, as being two or three years her junior, and married early. He celebrated her later on as "Lizbie Browne".' Hardy was in love with love. All his novels and most of his poems are concerned with it.

Faintheart in a Railway Train

AT nine in the morning there passed a church,
At ten there passed me by the sea,
At twelve a town of smoke and smirch,
At two a forest of oak and birch,
  And then, on a platform, she: [5]

A radiant stranger, who saw not me.
I said, 'Get out to her do I dare?'
But I kept my seat in my search for a plea,
And the wheels moved on. O could it but be
  That I had alighted there! [10]

A small incident but one common enough, and Hardy makes poetry out of it. Technical skill is seen in the ingenious rhyme scheme which calls for one rhyming sound to be used five times in ten lines.

At a Lunar Eclipse

THY shadow, Earth, from Pole to Central Sea,
Now steals along upon the Moon's meek shine
In even monochrome and curving line
Of imperturbable serenity.

How shall I link such sun-cast symmetry [5]
With the torn troubled form I know as thine,
That profile, placid as a brow divine,
With continents of moil and misery?

And can immense Mortality but throw
So small a shade, and Heaven's high human scheme [10]
Be hemmed within the coasts yon arc implies?

Is such the stellar gauge of earthly show,
Nation at war with nation, brains that teem,
Heroes, and women fairer than the skies?

J. O. Bailey identifies the eclipse as probably that of 16 July, 1860. Hardy, aged twenty, is already aware of how small the world is when seen against what he called 'the stupendous background of the stellar universe'. What has been described as his double vision enables Hardy to see both how important man's affairs are to himself and how unimportant when seen in relation to the vastness of space. We who have been able to see the world as a small ball photographed from a spacecraft can appreciate Hardy's thought all the better.

[8] **moil** turmoil, trouble.
[9] **Mortality** the human race.
[12] **stellar** of the stars.
[13] **teem** are fertile, prolific.

Neutral Tones

WE stood by a pond that winter day,
And the sun was white, as though chidden of God,
And a few leaves lay on the starving sod;
– They had fallen from an ash, and were grey.

Your eyes on me were as eyes that rove [5]
Over tedious riddles of years ago;
And some words played between us to and fro
On which lost the more by our love.

The smile on your mouth was the deadest thing
Alive enough to have strength to die; [10]
And a grin of bitterness swept thereby
    Like an ominous bird a-wing. . . .

Since then, keen lessons that love deceives,
And wrings with wrong, have shaped to me
Your face, and the God-curst sun, and a tree, [15]
    And a pond edged with greyish leaves.

1867

'Neutral Tones' has the effect of an etching in steel by a man trained in drawing the ruins of old churches, as Hardy was . . . The intensity of the etching and the particularity of its images suggest that it presents Hardy's actual experience.' (Prof. J. O. Bailey) It is part of Hardy's technique to use the natural background to support the mood of the poem. Here the relationship between the lovers is heightened by the description of the sombre scene. The ash tree, for example, should be a symbol of happiness, as it was once thought to be, but the happiness has passed away and its few remaining leaves lie on the ground and are grey.

In a Eweleaze near Weatherbury

THE years have gathered greyly
    Since I danced upon this leaze
With one who kindled gaily
    Love's fitful ecstasies!
But despite the term as teacher, [5]
    I remain what I was then
In each essential feature
    Of the fantasies of men.

Yet I note the little chisel
    Of never-napping Time, [10]
Defacing wan and grizzel
    The blazon of my prime.

When at night he thinks me sleeping
  I feel him boring sly
Within my bones, and heaping [15]
  Quaintest pains for by-and-by.

Still, I'd go the world with Beauty,
  I would laugh with her and sing,
I would shun divinest duty
  To resume her worshipping. [20]
But she'd scorn my brave endeavour,
  She would not balm the breeze
By murmuring 'Thine for ever!'
  As she did upon this leaze.

1890

This may be a recollection by Hardy of his romance of the late 1860s with Tryphena Sparks, who died in 1890. Weatherbury is Hardy's name for Puddletown where Tryphena lived at the time of the romance. It is about two miles from Hardy's cottage at Higher Bockhampton.

[11] **wan and grizzel** paleness and greyness.
[12] **blazon** sign or manifestation.
[16] **Quaintest** cunningest.

Great Things

SWEET cyder is a great thing,
  A great thing to me,
Spinning down to Weymouth town
  By Ridgeway thirstily,
And maid and mistress summoning, [5]
  Who tend the hostelry:
O cyder is a great thing,
  A great thing to me!

The dance it is a great thing,
  A great thing to me, [10]
With candles lit and partners fit
  For night-long revelry;
And going home when day-dawning
  Peeps pale upon the lea:
O dancing is a great thing, [15]
  A great thing to me!

Love is, yea, a great thing,
  A great thing to me,
When, having drawn across the lawn
  In darkness silently, [20]
A figure flits like one a-wing
  Out from the nearest tree:
O love is, yes, a great thing,
  A great thing to me!

Will these be always great things, [25]
  Great things to me? . . .
Let it befall that One will call,
  'Soul, I have need of thee':
What then? Joy-jaunts, impassioned flings,
  Love, and its ecstasy, [30]
Will always have been great things,
  Great things to me!

'The same sensibility that made Hardy so acutely susceptible to life's sorrows made him also exquisitely responsive to its joys.' (David Cecil) The sheer exuberance felt by Hardy comes through in the lyrical flow of the poem which is so technically accomplished that one rhyming sound is used effortlessly in alternate lines right through the poem.

[3] **Spinning** Hardy loved cycling.
[4] **Ridgeway** the name given to the old road which runs over the downs between Dorchester and Weymouth which are about seven miles apart.

'When I Set Out for Lyonnesse'
(1870)

WHEN I set out for Lyonnesse,
    A hundred miles away,
    The rime was on the spray,
And starlight lit my lonesomeness
When I set out for Lyonnesse [5]
    A hundred miles away.

What would bechance at Lyonnesse
    While I should sojourn there
    No prophet durst declare,
Nor did the wisest wizard guess [10]
What would bechance at Lyonnesse
    While I should sojourn there.

When I came back from Lyonnesse
    With magic in my eyes,
    All marked with mute surmise [15]
My radiance rare and fathomless,
When I came back from Lyonnesse
    With magic in my eyes!

In March 1870 Hardy travelled to St Juliot in Cornwall in order to plan the restoration of the parish church. There he met Emma Gifford, with whom he fell in love and eventually married in 1874. In the *Life* (p. 75), we read, ' "Lyonnesse" . . . was hailed . . . as his sweetest lyric, an opinion from which he (Hardy) himself did not dissent.'

[1] **Lyonnesse** Hardy's name for Cornwall.
[3] **rime . . . spray** Frost was on the small twigs of the shrubs and trees.
[10] **wizard** Cornwall was associated with magic and magicians.

The Sun on the Bookcase
(Student's Love-song: 1870)

ONCE more the cauldron of the sun
Smears the bookcase with winy red,
And here my page is, and there my bed,
And the apple-tree shadows travel along.
Soon their intangible track will be run, [5]
        And dusk grow strong
        And they have fled.

Yes: now the boiling ball is gone,
And I have wasted another day. . . .
But wasted – *wasted*, do I say? [10]
Is it a waste to have imaged one
Beyond the hills there, who, anon,
        My great deeds done,
        Will be mine alway?

'We Sat at the Window'
(Bournemouth, 1875)

WE sat at the window looking out,
And the rain came down like silken strings
That Swithin's day. Each gutter and spout
Babbled unchecked in the busy way
        Of witless things: [5]
Nothing to read, nothing to see
Seemed in that room for her and me
        On Swithin's day.

We were irked by the scene, by our own selves; yes,
For I did not know, nor did she infer [10]
How much there was to read and guess
By her in me, and to see and crown

By me in her.
Wasted were two souls in their prime,
And great was the waste, that July time [15]
When the rain came down.

Hardy and Emma, now his wife, visited Bournemouth in July 1875. This poem seems to describe the beginnings of that division between them which later led to so much unhappiness. It was not published until 1917 and may have been written many years after the incident.

[3] **Swithin's day** 15 July. There is an old belief that if it rains on July 15 it will rain on the next forty days. The symbolic significance of the date should be noticed.

A January Night
(1879)

THE rain smites more and more,
The east wind snarls and sneezes;
Through the joints of the quivering door
The water wheezes.

The tip of each ivy-shoot [5]
Writhes on its neighbour's face;
There is some hid dread afoot
That we cannot trace.

Is it the spirit astray
Of the man at the house below [10]
Whose coffin they took in today?
We do not know.

'The poem . . . relates to an incident . . . which occurred here at Tooting, where they (Hardy and his wife) seemed to begin to feel that "there had passed away a glory from the earth". And it was in this house that their troubles began.' (The *Life*, p. 124) 'This is a very characteristic poem, in that all we are allowed to *know* is the substantive situation – the wind, the rain, and the writhing ivy. The dread is what we do not and cannot know, the forces or the emptiness behind the actual.' (Samuel Hynes)

The Last Signal

11 (October, 1886)

A MEMORY OF WILLIAM BARNES

SILENTLY I footed by an uphill road
That led from my abode to a spot yew-boughed;
Yellowly the sun sloped low down to westward,
And dark was the east with cloud.

Then, amid the shadow of that livid sad east, [5]
Where the light was least, and a gate stood wide,
Something flashed the fire of the sun that was facing it,
Like a brief blaze on that side.

Looking hard and harder I knew what it meant –
The sudden shine sent from the livid east scene; [10]
It meant the west mirrored by the coffin of my friend there,
Turning to the road from his green,

To take his last journey forth – he who in his prime
Trudged so many a time from that gate athwart the land!
Thus a farewell to me he signalled on his grave-way, [15]
As with a wave of his hand.

Winterborne-Came Path

William Barnes had been a schoolmaster, poet and clergyman. His poems in the Dorset dialect are very fine indeed and were much admired by Hardy who had known Barnes for many years. At the end of his life Barnes became vicar of Winterborne-Came which was close to Hardy's house at Max Gate. As Hardy walked across the fields to attend Barnes's funeral the sun flashed on the coffin being carried to the church. Out of this incident Hardy wrote his poem. It is typical of Hardy's brilliant craftsmanship that he was able to pay a tribute to the dead man by using Welsh poetical techniques which Barnes had used. These included the use of internal rhymes (e.g. road/abode, east/least) and repetitive consonantal patterns (e.g. in line 3: LLSNSLLNS).

Shelley's Skylark

(The neighbourhood of Leghorn: March 1887)

SOMEWHERE afield here something lies
In Earth's oblivious eyeless trust
That moved a poet to prophecies –
A pinch of unseen, unguarded dust:

The dust of the lark that Shelley heard, [5]
And made immortal through times to be; –
Though it only lived like another bird,
And knew not its immortality:

Lived its meek life; then, one day, fell –
A little ball of feather and bone; [10]
And how it perished, when piped farewell,
And where it wastes, are alike unknown.

Maybe it rests in the loam I view,
Maybe it throbs in a myrtle's green,
Maybe it sleeps in the coming hue [15]
Of a grape on the slopes of yon inland scene.

Go find it, faeries, go and find
That tiny pinch of priceless dust,
And bring a casket silver-lined,
And framed of gold that gems encrust; [20]

And we will lay it safe therein,
And consecrate it to endless time ·
For it inspired a bard to win
Ecstatic heights in thought and rhyme.

During a holiday in Italy in 1887 Hardy and his wife visited Leghorn, near which the great English Romantic poet, Shelley (1792–1822) wrote his famous 'Ode to a Skylark'. Hardy was a lover of Shelley's poetry and cleverly combines a tribute to the poet with an expression of his love for animals, a love which we find repeatedly in Hardy's poetry.

At the Pyramid of Cestius near the Graves of Shelley and Keats
(1887)

WHO, then, was Cestius,
And what is he to me? –
Amid thick thoughts and memories multitudinous
One thought alone brings he.

I can recall no word [5]
Of anything he did;
For me he is a man who died and was interred
To leave a pyramid

Whose purpose was exprest
Not with its first design, [10]
Nor till, far down in Time, beside it found their rest
Two countrymen of mine.

Cestius in life, maybe,
Slew, breathed out threatening;
I know not. This I know: in death all silently [15]
He does a finer thing,

In beckoning pilgrim feet
With marble finger high
To where, by shadowy wall and history-haunted street,
Those matchless singers lie. . . . [20]

– Say, then, he lived and died
That stones which bear his name
Should mark, through Time, where two immortal Shades abide;
It is an ample fame.

The two great English Romantic poets, Shelley (1792–1822) and Keats (1795–1821), both died in Italy and are buried in Rome. Hardy visited their graves in 1887 and wrote this poem as a tribute to them. It grows out of the irony that the proud Cestius – an eminent Roman who died about 30 BC – is now forgotten in spite of his memorial pyramid, which now serves a far better purpose, that of indicating where the two poets lie.

The Division

RAIN on the windows, creaking doors,
With blasts that besom the green,
And I am here, and you are there,
And a hundred miles between!

O were it but the weather, Dear, [5]
O were it but the miles
That summed up all our severance,
There might be room for smiles.

But that thwart thing betwixt us twain,
Which hides, yet reappears, [10]
Is more than distance, Dear, or rain,
And longer than the years!

This love poem with its hint of something more than mere miles separating the lovers may refer to the intense gap which by 1893 had developed between Hardy and his wife, or it may refer to the emotional involvement he had at that time for a society woman called Mrs Henniker. She had used Hardy to further her literary career and they wrote a short story together. It has been variously suggested that the 'thwart thing' is his wife's mental instability and his wife's jealousy of Mrs Henniker. Whatever the truth, it remains a moving poem of separation and loss.

[2] **besom** sweep violently.
[9] **thwart** adverse, obstructive.

A Broken Appointment

You did not come,
And marching Time drew on, and wore me numb. –
Yet less for loss of your dear presence there
Than that I thus found lacking in your make
That high compassion which can overbear [5]
Reluctance for pure lovingkindness' sake
Grieved I, when, as the hope-hour stroked its sum,
You did not come.

You love not me,
And love alone can lend you loyalty; [10]
– I know and knew it. But, unto the store
Of human deeds divine in all but name,
Was it not worth a little hour or more
To add yet this: Once you, a woman, came
To soothe a time-torn man; even though it be [15]
You love not me?

It is generally thought that this is another poem written in the 1890s and referring to Mrs Henniker. Like most of Hardy's poems it is at the same time both highly personal and universal, that is to say it says something to all lovers who have at one time or another been 'stood up' and disappointed. For Hardy 'lovingkindness' was all important and it is one of his favourite words. Notice how much meaning and feeling he is able to get into 'time-torn'.

[4] **make** make-up, character.

The Impercipient
(At a Cathedral Service)

THAT with this bright believing band
I have no claim to be,
That faiths by which my comrades stand
Seem fantasies to me,
And mirage-mists their Shining Land, [5]
Is a strange destiny.

Why thus my soul should be consigned
To infelicity,
Why always I must feel as blind
To sights my brethren see, [10]
Why joys they've found I cannot find,
Abides a mystery.

Since heart of mine knows not that ease
    Which they know; since it be
That He who breathes All's Well to these [15]
    Breathes no All's-Well to me,
My lack might move their sympathies
    And Christian charity!

I am like a gazer who should mark
    An inland company [20]
Standing upfingered, with, 'Hark! hark!
    The glorious distant sea!'
And feel, 'Alas, 'tis but yon dark
    And wind-swept pine to me!'

Yet I would bear my shortcomings [25]
    With meet tranquillity,
But for the charge that blessed things
    I'd liefer not have be.
O, doth a bird deprived of wings
    Go earth-bound wilfully! [30]

Enough. As yet disquiet clings
    About us. Rest shall we.

When this poem first appeared in *Wessex Poems* in 1898 it was illustrated by a drawing of Salisbury Cathedral drawn by Hardy himself. Hardy's loss of faith and the attacks on him because of what were described as his 'irreligious novels' caused him great distress and an acute sense of loss which is felt very powerfully here even though Hardy's tone is gentle and wistful. The final emphasis is on the fact that all of us – believers and unbelievers – are faced with the same eternal rest.

Impercipient one who cannot perceive or believe.
[8] **infelicity** unhappiness.
[26] **meet** suitable.
[28] **liefer** rather.

The Going of the Battery

WIVES' LAMENT

(2 November, 1899)

O IT was sad enough, weak enough, mad enough –
Light in their loving as soldiers can be –
First to risk choosing them, leave alone losing them
Now, in far battle, beyond the South Sea! . . .

– Rain came down drenchingly; but we unblenchingly [5]
Trudged on beside them through mirk and through mire,
They stepping steadily – only too readily! –
Scarce as if stepping brought parting-time nigher.

Great guns were gleaming there, living things seeming there,
Cloaked in their tar-cloths, upmouthed to the night; [10]
Wheels wet and yellow from axle to felloe,
Throats blank of sound, but prophetic to sight.

Gas-glimmers drearily, blearily, eerily
Lit our pale faces outstretched for one kiss,
While we stood prest to them, with a last quest to them [15]
Not to court perils that honour could miss.

Sharp were those sighs of ours, blinded these eyes of ours,
When at last moved away under the arch
All we loved. Aid for them each woman prayed for them,
Treading back slowly the track of their march. [20]

Some one said: 'Nevermore will they come: evermore
Are they now lost to us.' O it was wrong!
Though may be hard their ways, some Hand will guard their ways,
Bear them through safely, in brief time or long.

– Yet, voices haunting us, daunting us, taunting us, [25]
Hint in the night-time when life beats are low
Other and graver things. . . . Hold we to braver things,
Wait we, in trust, what Time's fulness shall show.

When this poem first appeared in *The Graphic* magazine it bore the note: 'November 2, 1899. Late at night, in rain and in darkness, the 73rd Battery, Royal Field Artillery, left Dorchester Barracks for the War in South Africa, marching on foot to the railway station, where their guns were already entrained.' Hardy was so moved by the pathos of the scene that he stayed up until the early hours of the morning. His 'lovingkindness' is seen in the way in which he is able to depict and share the sorrow of the wives. 'The rhythm of the poem, with its internal rhymes . . . represents the swinging march of the soldiers mingled with the equally rhythmical lament of the wives.' (J. O. Bailey)

[11] **felloe** rim.

Drummer Hodge

THEY throw in Drummer Hodge, to rest
    Uncoffined – just as found:
His landmark is a kopje-crest
    That breaks the veldt around;
And foreign constellations west [5]
    Each night above his mound.

Young Hodge the Drummer never knew –
    Fresh from his Wessex home –
The meaning of the broad Karoo,
    The Bush, the dusty loam, [10]
And why uprose to nightly view
    Strange stars amid the gloam.

Yet portion of that unknown plain
    Will Hodge for ever be;
His homely Northern breast and brain [15]
    Grow to some Southern tree,
And strange-eyed constellations reign
    His stars eternally.

One of the greatest of all war poems. It was originally called 'The Drummer Boy' and the change of title is important. 'Hodge' was a nickname given to the so-called country yokel, but Hardy objected strongly to its use because, as an individualist, who loved and understood country-people, he knew that every 'Hodge' was an individual. But in war individuals don't matter, the drummer boy becomes Drummer Hodge, mere cannon-fodder, and his body is *thrown*, uncoffined, just as found, into the ground. There is no false heroism – just a realistic, honest description which through its restraint brings out the sheer horror, waste, and pity of war.

[3] **kopje** a small hill.
[4] **veldt** open grassland.
[9] **Karoo** an arid tableland.
[10] **Bush** shrubby, uncultivated country.

The Man He Killed

'HAD he and I but met
By some old ancient inn,
We should have sat us down to wet
Right many a nipperkin!

'But ranged as infantry, [5]
And staring face to face,
I shot at him as he at me,
And killed him in his place.

'I shot him dead because –
Because he was my foe, [10]
Just so: my foe of course he was;
That's clear enough; although

'He thought he'd 'list, perhaps,
Off-hand like – just as I –
Was out of work – had sold his traps – [15]
No other reason why.

'Yes; quaint and curious war is!
You shoot a fellow down
You'd treat if met where any bar is,
Or help to half-a-crown.' [20]

1902

[4] **nipperkin** a small measure of drink.
[20] **half-a-crown** a unit of money, equal to 12½p.

The Darkling Thrush

I LEANT upon a coppice gate
    When Frost was spectre-grey,
And Winter's dregs made desolate
    The weakening eye of day.
The tangled bine-stems scored the sky [5]
    Like strings of broken lyres,
And all mankind that haunted nigh
    Had sought their household fires.

The land's sharp features seemed to be
    The Century's corpse outleant, [10]
His crypt the cloudy canopy,
    The wind his death-lament.
The ancient pulse of germ and birth
    Was shrunken hard and dry,
And every spirit upon earth [15]
    Seemed fervourless as I.

At once a voice arose among
    The bleak twigs overhead
In a full-hearted evensong
    Of joy illimited; [20]
An aged thrush, frail, gaunt, and small,
    In blast-beruffled plume,
Had chosen thus to fling his soul
    Upon the growing gloom.

So little cause for carolings [25]
    Of such ecstatic sound
Was written on terrestrial things
      Afar or nigh around,
That I could think there trembled through
    His happy good-night air [30]
Some blessed Hope, whereof he knew
    And I was unaware.

31 December, 1900

'It is the way Hardy is driven, against all the evidence of his senses, to accept the possibility, however faint, of "some blessed Hope", the perfect balance between reason and emotion in the poem, the contrast between the deathless beauty of the music and the deathstruck setting for it, with, perhaps above all else, the hesitance, the shyness almost, with which Hardy advances his infinitely precarious conclusion, that makes this a great poem. Whether we accept the "Hope" as justified or not, the image of the thrush remains, gallant and unconquerable, an image of man himself "slighted but enduring".' (Trevor Johnson) Hardy wrote this at a time of despair. His marriage was at its lowest ebb, his religious faith lost. He was now sixty and so much of what he had loved was passing away. Technically the poem is worthy of very close study. Note particularly the setting of the scene, the way in which the landscape becomes a symbol of the dead century, and the use of religious imagery.

Darkling shrouded in darkness.
[4] **eye** i.e. the sun.
[5] **bine-stems** the stems of last year's climbing plants like the woodbine.
[6] **lyres** ancient stringed musical instruments.
[10] **outleant** possibly laid out like a corpse.
[20] **illimited** unlimited.

To an Unborn Pauper Child

BREATHE not, hid Heart: cease silently,
And though thy birth-hour beckons thee,
Sleep the long sleep:
The Doomsters heap
Travails and teens around us here, [5]
And Time-wraiths turn our songsingings to fear.

Hark, how the peoples surge and sigh,
And laughters fail, and greetings die:
Hopes dwindle; yea,
Faiths waste away, [10]
Affections and enthusiasms numb;
Thou canst not mend these things if thou dost come.

Had I the ear of wombèd souls
Ere their terrestrial chart unrolls,
And thou wert free [15]
To cease, or be,
Then would I tell thee all I know,
And put it to thee: Wilt thou take Life so?

Vain vow! No hint of mine may hence
To theeward fly: to thy locked sense [20]
Explain none can
Life's pending plan:
Thou wilt thy ignorant entry make
Though skies spout fire and blood and nations quake.

Fain would I, dear, find some shut plot [25]
Of earth's wide wold for thee, where not
One tear, one qualm,
Should break the calm.
But I am weak as thou and bare;
No man can change the common lot to rare. [30]

Must come and bide. And such are we –
Unreasoning, sanguine, visionary –
That I can hope
Health, love, friends, scope
In full for thee; can dream thou'lt find [35]
Joys seldom yet attained by humankind!

Hardy's manuscript adds, '"She must go to the Union-House (poorhouse) to have her baby." Casterbridge Petty Sessions.' Again a poem, made of concern to us all by Hardy's compassion, sympathy, and humility, grows out of an incident that he probably witnessed in the Dorchester (Casterbridge) Magistrates' Court. Again there is the hope that in spite of the evidence the unborn child of the penniless woman will find happiness in life. Written probably about 1900 this poem shows Hardy at his brilliant best. Note particularly the love with which he speaks to the child and how in stanza 5 almost every word is a monosyllable.

[4] **Doomsters** Fates, i.e. those who pronounce doom on us.
[5] **teens** troubles, sorrows.
[6] **Time-wraiths** spirits who make us aware of the passing of time, of the coming of old age.
[25] **shut** secluded.
[26] **wold** open country.

Shut out that Moon

CLOSE up the casement, draw the blind,
Shut out that stealing moon,
She wears too much the guise she wore
Before our lutes were strewn
With years-deep dust, and names we read [5]
On a white stone were hewn.

Step not forth on the dew-dashed lawn
To view the Lady's Chair,
Immense Orion's glittering form,
The Less and Greater Bear: [10]
Stay in; to such sights we were drawn
When faded ones were fair.

Brush not the bough for midnight scents
That come forth lingeringly,
And wake the same sweet sentiments [15]
They breathed to you and me
When living seemed a laugh, and love
All it was said to be.

Within the common lamp-lit room
Prison my eyes and thought; [20]
Let dingy details crudely loom,
Mechanic speech be wrought:
Too fragrant was Life's early bloom,
Too tart the fruit it brought!

(Written in 1904)

By 1904 Hardy's life seemed to have gone sour and in the poems published in *Time's Laughingstocks* in 1909 there is a noticeable vein of tartness, even cynicism.

[2] **stealing** the moon has quietly stolen away youth and beauty and love.
[8] **Lady's Chair** a famous group of stars, otherwise known as Cassiopeia.
[9] **Orion** a large and brilliant group of stars imagined as representing a hunter with belt and sword.
[10] **Less and Greater Bear** two more groups of stars.
[22] **Mechanic . . . wrought** let talk be merely mechanical, lacking all vitality and spontaneity.

After the Last Breath

(J.H. 1813–1904)

THERE'S no more to be done, or feared, or hoped;
None now need watch, speak low, and list, and tire;
No irksome crease outsmoothed, no pillow sloped
Does she require.

Blankly we gaze. We are free to go or stay; [5]
Our morrow's anxious plans have missed their aim;
Whether we leave tonight or wait till day
Counts as the same.

The lettered vessels of medicaments
Seem asking wherefore we have set them here; [10]
Each palliative its silly face presents
As useless gear.

And yet we feel that something savours well;
We note a numb relief withheld before;
Our well-beloved is prisoner in the cell [15]
Of Time no more.

We see by littles now the deft achievement
Whereby she has escaped the Wrongers all,
In view of which our momentary bereavement
Outshapes but small. [20]

1904

Hardy's mother, Jemima, died on 1 April 1904. The moment after death is movingly and vividly portrayed and made universal by 'momentary bereavement' with its implication that we shall all follow into death those separated from us by what will eventually seem no more than a moment.

[9] **lettered . . . medicaments** medicine bottles with writing on them.
[11] **palliatives** the various aids to make life easier for the patient.
[18] **Wrongers** Hardy probably means that his mother has escaped from the trials and infirmities which surround us all.

The Convergence of the Twain

(Lines on the loss of the *Titanic*)

IN a solitude of the sea
Deep from human vanity,
And the Pride of Life that planned her, stilly couches she.

Steel chambers, late the pyres
Of her salamandrine fires, [5]
Cold currents thrid, and turn to rhythmic tidal lyres.

Over the mirrors meant
To glass the opulent
The sea-worm crawls – grotesque, slimed, dumb, indifferent.

Jewels in joy designed [10]
To ravish the sensuous mind
Lie lightless, all their sparkles bleared and black and blind.

Dim moon-eyed fishes near
Gaze at the gilded gear
And query: 'What does this vaingloriousness down here?' [15]

Well: while was fashioning
This creature of cleaving wing,
The Immanent Will that stirs and urges everything

Prepared a sinister mate
For her – so gaily great – [20]
A Shape of Ice, for the time far and dissociate.

And as the smart ship grew
In stature, grace, and hue,
In shadowy silent distance grew the Iceberg too.

Alien they seemed to be: [25]
No mortal eye could see
The intimate welding of their later history,

Or sign that they were bent
By paths coincident
On being anon twin halves of one august event, [30]

Till the Spinner of the Years
Said 'Now!' And each one hears,
And consummation comes, and jars two hemispheres.

On 15 April 1912 the liner, *Titanic*, on her maiden voyage across the Atlantic, hit an iceberg and sank with the loss of over 1,500 lives. The 'Pride of Life' refers to the enormous luxury with which the ship was fitted out and the belief that she was unsinkable. Hardy was moved by two aspects of the sinking – the irony that the unsinkable should be sunk and the realisation that once again it had been shown that man, for all his greatness, is at the mercy of an unknown fate. It never does to be too proud. The poem is cleverly structured. The stanzas are designed to look something like a ship, and the opening stanzas are built on the contrast between what was (the first two lines of each stanza) and what is (the third line). The opulent mirrors are now crawled over by sea-worms at the bottom of the sea. The first five stanzas deal with the building of the ship, the next three stanzas with the growing of the iceberg, the final three stanzas their 'marriage'.

[4] **pyres** the places where the fires of the ship's engines burn.
[5] **salamandrine** the lizard-like animal which is called the salamander is said in legend to be able to endure fire.

[6] **thrid** thread their way through.
[8] **glass** reflect.
[17] **cleaving wing** The ship cleaves or parts the sea as a bird's wings do the air.
[18] **The Immanent Will** Hardy believed that there was an unconscious purpose – the Immanent Will – working itself out through history and indifferent to our welfare.
[30] **anon** soon.
[31] **Spinner of the Years** another of Hardy's names for Fate or whatever unknown force controls our lives. In olden times it was believed that there were three old women who spun, measured and cut the thread of men's lives.

At Tea

THE kettle descants in a cosy drone,
And the young wife looks in her husband's face,
And then at her guest's, and shows in her own
Her sense that she fills an envied place;
And the visiting lady is all abloom, [5]
And says there was never so sweet a room.

And the happy young housewife does not know
That the woman beside her was first his choice,
Till the fates ordained it could not be so. . . .
Betraying nothing in look or voice [10]
The guest sits smiling and sips her tea,
And he throws her a stray glance yearningly.

In 1911 Hardy published in a magazine a series of short poems to which he gave the general title 'Satires of Circumstance'. This and the next two poems are three of that series. They are called satires because Hardy is using irony, in each poem things are not what they seem or what was intended. Hardy himself described them as being 'rather brutal' and the humour does not hide a certain sick quality about them which may tell the reader something about the state of Hardy's mind at that time. It is significant that most of them are about unhappy love affairs or marriages that have gone wrong.

By Her Aunt's Grave

'SIXPENCE a week,' says the girl to her lover,
'Aunt used to bring me, for she could confide
In me alone, she vowed. 'Twas to cover
The cost of her headstone when she died.
And that was a year ago last June; [5]
I've not yet fixed it. But I must soon.'

'And where is the money now, my dear?'
'O, snug in my purse . . . Aunt was *so* slow
In saving it – eighty weeks, or near.' . . .
'Let's spend it,' he hints. 'For she won't know. [10]
There's a dance tonight at the Load of Hay.'
She passively nods. And they go that way.

[11] **Load of Hay** the name of a public house where people go to drink and dance.

In the Cemetery

'You see those mothers squabbling there?'
Remarks the man of the cemetery.
'One says in tears, "*'Tis mine lies here!*"
Another, "*Nay, mine, you Pharisee!*"
Another, "*How dare you move my flowers* [5]
And put your own on this grave of ours!"
But all their children were laid therein
At different times, like sprats in a tin.

'And then the main drain had to cross,
And we moved the lot some nights ago, [10]
And packed them away in the general foss
With hundreds more. But their folks don't know,
And as well cry over a new-laid drain
As anything else, to ease your pain!'

[4] **Pharisee** A Pharisee was a very strict member of the Jewish faith and the name became associated with hypocrisy and thus became a term of contempt.
[11] **foss** trench.

The Going

WHY did you give no hint that night
That quickly after the morrow's dawn,
And calmly, as if indifferent quite,
You would close your term here, up and be gone
    Where I could not follow [5]
    With wing of swallow
To gain one glimpse of you ever anon!

    Never to bid good-bye,
    Or lip me the softest call,
Or utter a wish for a word, while I [10]
Saw morning harden upon the wall,
    Unmoved, unknowing
    That your great going
Had place that moment, and altered all.

Why do you make me leave the house [15]
And think for a breath it is you I see
At the end of the alley of bending boughs
Where so often at dusk you used to be;
    Till in darkening dankness
    The yawning blankness [20]
Of the perspective sickens me!

    You were she who abode
    By those red-veined rocks far West,
You were the swan-necked one who rode
Along the beetling Beeny Crest, [25]
    And, reining nigh me,
    Would muse and eye me,
While Life unrolled us its very best.

Why, then, latterly did we not speak,
Did we not think of those days long dead, [30]
And ere your vanishing strive to seek
That time's renewal? We might have said,
'In this bright spring weather
We'll visit together
Those places that once we visited.' [35]

Well, well! All's past amend,
Unchangeable. It must go.
I seem but a dead man held on end
To sink down soon. . . . O you could not know
That such swift fleeing [40]
No soul foreseeing –
Not even I – would undo me so!

December 1912

In November 1912 Hardy's wife, Emma, died. It was forty-two years since he had visited 'Lyonnesse' and fallen in love with her. The romance had begun rapturously but the marriage had grown increasingly unhappy. Immediately after Emma's death Hardy was overcome with sorrow and remorse, visited Cornwall again, and wrote some of his finest poetry about his deep sense of loss and regret for the wasted years. This and subsequent poems are all concerned with Emma's death and Hardy's sadness. What is impressive is the variety of approach, the technical skill with which he depicts his grief, and, above all, the honesty and sincerity which universalises his experience so that 'What begins with an obscure private hurt ends with the common wound of experience' (Irving Howe). 'The Going' is about Emma's death which was sudden and unexpected. The 'alley of bending boughs' was an avenue in Max Gate garden where Emma used often to tend her flowers in the evenings. 'Beeny Crest' takes us back forty years to Cornwall. Beeny is a magnificent cliff near Emma's old home at St Juliot where Emma used to ride her horse. It is surprising that after their marriage they never revisited Cornwall together.

[14] **had place** took place.
[23] **red-veined** reddish-brown quartz veins found in the Cornish rock.
[25] **beetling** jutting out.
[26] **reining** stopping the horse by the reins.

The Haunter

HE does not think that I haunt here nightly:
How shall I let him know
That whither his fancy sets him wandering
I, too, alertly go? –
Hover and hover a few feet from him [5]
Just as I used to do,
But cannot answer the words he lifts me –
Only listen thereto!

When I could answer he did not say them:
When I could let him know [10]
How I would like to join in his journeys
Seldom he wished to go.
Now that he goes and wants me with him
More than he used to do,
Never he sees my faithful phantom [15]
Though he speaks thereto.

Yes, I companion him to places
Only dreamers know,
Where the shy hares print long paces,
Where the night rooks go; [20]
Into old aisles where the past is all to him,
Close as his shade can do,
Always lacking the power to call to him,
Near as I reach thereto!

What a good haunter I am, O tell him! [25]
Quickly make him know
If he but sigh since my loss befell him
Straight to his side I go.
Tell him a faithful one is doing
All that love can do [30]
Still that his path may be worth pursuing,
And to bring peace thereto.

The speaker is the ghost of Emma. Hardy uses a favourite device – a linked rhyming scheme. The rhyme words of the even lines of the first stanza are reproduced exactly in each of the following stanzas.

[21] **old aisles** Hardy loved old churches.

The Walk

YOU did not walk with me
Of late to the hill-top tree
By the gated ways,
As in earlier days;
You were weak and lame, [5]
So you never came,
And I went alone, and I did not mind,
Not thinking of you as left behind.

I walked up there today
Just in the former way; [10]
Surveyed around
The familiar ground
By myself again:
What difference, then?
Only that underlying sense [15]
Of the look of a room on returning thence.

The Voice

WOMAN much missed, how you call to me, call to me,
Saying that now you are not as you were
When you had changed from the one who was all to me,
But as at first, when our day was fair.

Can it be you that I hear? Let me view you, then, [5]
Standing as when I drew near to the town
Where you would wait for me: yes, as I knew you then,
Even to the original air-blue gown!

Or is it only the breeze, in its listlessness
Travelling across the wet mead to me here, [10]
You being ever dissolved to existlessness,
Heard no more again far or near?

Thus I; faltering forward,
Leaves around me falling,
Wind oozing thin through the thorn from norward, [15]
And the woman calling.

December 1912

'Hardy's great poems, as a rule, start immediately out of his own remembered past, and are particular evocations of utter loss, the blindness of chance, the poignancy of love and its helplessness, and the cruelty of time. Such poems are 'After a Journey', 'The Voice' . . .' (F. R. Leavis) Note how the break-down of rhythm in the last stanza conveys the feeling of Hardy's heart-broken state as he goes 'faltering forward'.

At Castle Boterel

As I drive to the junction of lane and highway,
And the drizzle bedrenches the waggonette,
I look behind at the fading byway,
And see on its slope, now glistening wet,
Distinctly yet [5]

Myself and a girlish form benighted
In dry March weather. We climb the road
Beside a chaise. We had just alighted
To ease the sturdy pony's load
When he sighed and slowed. [10]

What we did as we climbed, and what we talked of
Matters not much, nor to what it led, –
Something that life will not be balked of
Without rude reason till hope is dead,
And feeling fled. [15]

It filled but a minute. But was there ever
  A time of such quality, since or before,
In that hill's story? To one mind never,
  Though it has been climbed, foot-swift, foot-sore,
      By thousands more. [20]

Primaeval rocks form the road's steep border,
  And much have they faced there, first and last,
Of the transitory in Earth's long order;
  But what they record in colour and cast
      Is – that we two passed. [25]

And to me, though Time's unflinching rigour,
  In mindless rote, has ruled from sight
The substance now, one phantom figure
  Remains on the slope, as when that night
      Saw us alight. [30]

I look and see it there, shrinking, shrinking,
  I look back at it amid the rain
For the very last time; for my sand is sinking,
  And I shall traverse old love's domain
      Never again. [35]

March 1913

Castle Boterel is Hardy's name for Boscastle, a small Cornish town about a mile from St Juliot where Emma lived when Hardy first met her. Again the scene is set by the precise but imaginative use of detail remembered by Hardy over a period of forty years. What happened in the 'minute' – possibly a declaration of love – is left to the imagination. For Hardy the importance lies in the fact that it happened and is still remembered in spite of Time's hardness in removing the substance of what happened.

[6] **benighted** overtaken by night.
[21] **Primaeval** so old that they go back to the beginnings of the world.
[26–7] **Time's . . . rote** Time is seen as both a stern taskmaster who will stop for no one and as a mindless, repetitive machine.

After a Journey

HERETO I come to view a voiceless ghost;
Whither, O whither will its whim now draw me.
Up the cliff, down, till I'm lonely, lost,
And the unseen waters' ejaculations awe me.
Where you will next be there's no knowing, [5]
Facing round about me everywhere,
With your nut-coloured hair,
And grey eyes, and rose-flush coming and going.

Yes: I have re-entered your olden haunts at last;
Through the years, through the dead scenes I have tracked you; [10]
What have you now found to say of our past –
Scanned across the dark space wherein I have lacked you?
Summer gave us sweets, but autumn wrought division?
Things were not lastly as firstly well
With us twain, you tell? [15]
But all's closed now, despite Time's derision.

I see what you are doing: you are leading me on
To the spots we knew when we haunted here together,
The waterfall, above which the mist-bow shone
At the then fair hour in the then fair weather, [20]
And the cave just under, with a voice still so hollow
That it seems to call out to me from forty years ago,
When you were all aglow,
And not the thin ghost that I now frailly follow!

Ignorant of what there is flitting here to see, [25]
The waked birds preen and the seals flop lazily;
Soon you will have, Dear, to vanish from me,
For the stars close their shutters and the dawn whitens hazily.
Trust me, I mind not, though Life lours,
The bringing me here; nay, bring me here again! [30]
I am just the same as when
Our days were a joy, and our paths through flowers.

(*Written at Pentargan Bay*)

Hardy is still in Cornwall on his visit of March 1913, after the death of his wife. Pentargan Bay is less than a mile north-east of Boscastle and he and Emma would often have visited the Bay during their courtship in the 1870s. Of this poem F. R. Leavis has written, 'It is a poem that we recognise to have come directly out of life: it could, that is, have been written only by a man who had the experience of a life to remember back through. And recognising that, we recognise the rare quality of a man who can say with that truth "I am just the same", and the rare integrity that can so put the truth beyond question.'

[4] **ejaculations** Hardy is awed by the noise of the waters which sounds to him like sudden short exclamations. This provides a remarkably vivid description of the slap of the waves on the rocks.

Beeny Cliff
(March 1870–March 1913)

O THE opal and the sapphire of that wandering western sea,
And the woman riding high above with bright hair flapping free –
The woman whom I loved so, and who loyally loved me.

The pale mews plained below us, and the waves seemed far away
In a nether sky, engrossed in saying their ceaseless babbling
say, [5]
As we laughed light-heartedly aloft on that clear-sunned March
day.

A little cloud then cloaked us, and there flew an irised rain,
And the Atlantic dyed its levels with a dull misfeatured stain,
And then the sun burst out again, and purples prinked the main.

– Still in all its chasmal beauty bulks old Beeny to the sky, [10]
And shall she and I not go there once again now March is nigh,
And the sweet things said in that March say anew there by and
by?

What if still in chasmal beauty looms that wild weird western
shore,
The woman now is – elsewhere – whom the ambling pony bore,
And nor knows nor cares for Beeny, and will laugh there nevermore. [15]

This is another of the poems resulting from Hardy's pilgrimage to Cornwall in March 1913. As so often with Hardy the poem is built up on the contrast between the past and the present, and between the shortness of human life and the comparative permanence of nature as seen in old Beeny bulking to the sky.

[4] **mews** gulls.
plained Gulls make a cry which sounds rather like complaining. There may also be a deliberate pun on 'planed' which would also make sense.
[5] **nether** lower.
[7] **irised rain** a shower of rain containing a rainbow.
[9] **prinked** bedecked, ornamented.
[10] **chasmal** deeply fissured or cracked, and covering a great space of time.
[13] **weird** Cornwall was associated with magic and mystery.

The Frozen Greenhouse

(St Juliot)

'THERE was a frost
Last night!' she said,
'And the stove was forgot
When we went to bed,
And the greenhouse plants [5]
Are frozen dead!'

By the breakfast blaze
Blank-faced spoke she,
Her scared young look
Seeming to be [10]
The very symbol
Of tragedy.

The frost is fiercer
Than then today,
As I pass the place [15]
Of her once dismay,
But the greenhouse stands
Warm, tight, and gay,

While she who grieved
At the sad lot [20]
Of her pretty plants –
Cold, iced, forgot –
Herself is colder,
And knows it not.

Almost certainly written during or soon after Hardy's 1913 visit to St Juliot, the poem recalls an incident of the 1870s. There is a large greenhouse attached to the St Juliot Rectory. The dramatic opening with Emma speaking gives a feeling of genuine ness and actuality, as does the extreme simplicity of the language.

On a Discovered Curl of Hair

WHEN your soft welcomings were said,
This curl was waving on your head,
And when we walked where breakers dinned
It sported in the sun and wind,
And when I had won your words of grace [5]
It brushed and clung about my face.
Then, to abate the misery
Of absentness, you gave it me.

Where are its fellows now? Ah, they
For brightest brown have donned a grey, [10]
And gone into a caverned ark,
Ever unopened, always dark!

Yet this one curl, untouched of time,
Beams with live brown as in its prime,
So that it seems I even could now [15]
Restore it to the living brow
By bearing down the western road
Till I had reached your old abode.

February 1913

Hardy was fascinated by Emma's hair. He refers to it in both 'After a Journey' and 'Beeny Cliff'. Here he imagines himself talking to his dead wife three months after her death. He kept the lock of hair, which Emma had given him in the 1870s, in a green leather locket.

[11] **cavemed ark** coffin.
[17] **the western road** to Cornwall.

A Night in November

I MARKED when the weather changed,
And the panes began to quake,
And the winds rose up and ranged,
That night, lying half-awake.

Dead leaves blew into my room, [5]
And alighted upon my bed,
And a tree declared to the gloom
Its sorrow that they were shed.

One leaf of them touched my hand,
And I thought that it was you [10]
There stood as you used to stand,
And saying at last you knew!

(*?*) *1913*

[12] **at last you knew** i.e. in spite of all their differences he really loved her.

During Wind and Rain

THEY sing their dearest songs –
He, she, all of them – yea,
Treble and tenor and bass,
And one to play;
With the candles mooning each face. . . . [5]
Ah, no; the years O!
How the sick leaves reel down in throngs!

They clear the creeping moss –
Elders and juniors – aye,
Making the pathways neat [10]
And the garden gay;
And they build a shady seat. . . .
Ah, no; the years, the years;
See, the white storm-birds wing across!

They are blithely breakfasting all – [15]
Men and maidens – yea,
Under the summer tree,
With a glimpse of the bay,
While pet fowl come to the knee. . . .
Ah, no; the years O! [20]
And the rotten rose is ript from the wall.

They change to a high new house,
He, she, all of them – aye,
Clocks and carpets and chairs
On the lawn all day, [25]
And brightest things that are theirs. . . .
Ah, no; the years, the years;
Down their carved names the rain-drop ploughs.

Shortly before she died Emma Hardy wrote an account of her childhood in Plymouth and her meeting with Hardy in Cornwall, their falling in love, and marriage. She called her account *Some Recollections* and this poem is built upon those memories. Just after she died Hardy found and read her little book and was very moved by it. In addition to visiting Cornwall again, he also visited Plymouth and saw the house at 9 Bedford Terrace from which Emma as a child had been able to see the fishing boats in the bay. In the poem Hardy's sense of structure can be felt very strongly. In each stanza the first five lines describe an incident of the past, full of life and vigour, the sixth line provides a refrain, cleverly varied in stanzas 2 and 4, which brings change, age, and decay upon the scene, and then with a powerful last line Hardy shows the inevitability of Time's final victory and the oblivion that faces us all.

[4] **one to play** i.e. the piano.
[28] **Down . . . ploughs** At an early stage of the poem's composition this line read, 'On their chiselled names the lichen grows.'

Channel Firing

THAT night your great guns, unawares,
Shook all our coffins as we lay,
And broke the chancel window-squares,
We thought it was the Judgement-day

And sat upright. While drearisome [5]
Arose the howl of wakened hounds:
The mouse let fall the altar-crumb,
The worms drew back into the mounds,

The glebe cow drooled. Till God called, 'No;
It's gunnery practice out at sea [10]
Just as before you went below;
The world is as it used to be:

'All nations striving strong to make
Red war yet redder. Mad as hatters
They do no more for Christés sake [15]
Than you who are helpless in such matters.

'That this is not the judgement-hour
For some of them's a blessed thing,
For if it were they'd have to scour
Hell's floor for so much threatening. . . [20]

'Ha, ha. It will be warmer when
I blow the trumpet (if indeed
I ever do; for you are men,
And rest eternal sorely need).'

So down we lay again. 'I wonder, [25]
Will the world ever saner be,'
Said one, 'than when He sent us under
In our indifferent century!'

And many a skeleton shook his head.
'Instead of preaching forty year,' [30]
My neighbour Parson Thirdly said,
'I wish I had stuck to pipes and beer.'

Again the guns disturbed the hour,
Roaring their readiness to avenge,
As far inland as Stourton Tower, [35]
And Camelot, and starlit Stonehenge.

April 1914

The narrator is a dead man, the occasion is an imagined conversation between God and the dead in the coffins buried in the church who have been disturbed by the noise of gunnery practice out at sea. In the days of big battleships this sound could be heard for many miles. Hardy's tone is both humorous and serious. The idea of God as a kind of headmaster rebuking men for their folly is obviously comic, and yet his observation that the world shows itself to be as mad as ever by its attempt to make 'red war yet redder' is frighteningly serious. And less than six months after the writing of this prophetic poem the madmen were to go into yet another war, the 'reddest' of all. The poem ends with an ominous reference to three dead civilisations of the past – Stourton Tower commemorates the victory of King Alfred over the Danes, Camelot is the city of King Arthur, and Stonehenge a religious memorial of the distant past.

[3] **chancel** the eastern part of a church containing the choir and sanctuary.

[9] **glebe cow** cow grazing on church land.
drooled dribbled at the mouth.

1 Bronze head of Hardy, aged 83, by Maggie Richardson

2 (above) Emma Gifford, 1870, later Hardy's wife
3 (below) Hardy's drawing of his birthplace ('Domicilium', see p. 29)

4 St Juliot church, Cornwall

After a journey.

Hereto I come to interview a ghost;
Whither O whither will its whim now draw me?
Up the cliff, down, till I'm lonely, lost,
And the unseen waters' soliloquies awe me.
Where you will next be there's no knowing,
Facing round about me everywhere,
With your nut-coloured hair,
And grey eyes, and rose flush coming and going.

Yes: I have re-entered your olden haunts at last;
Through the years, through the dead scenes I have tracked;
What have you now found to say of our past—
Viewed across the dark space wherein I have lacked you?
Summer gave us sweets, but autumn wrought division?
Things were not lastly as firstly well
With us twain, you tell?
But all's soothed now, despite Time's derision.

5–6 Manuscript of 'After a Journey' (see p. 73)

I see what you are doing: you are leading me on
To the spots we knew when we haunted here together,
The waterfall, above which the mist-bow shone
At the then fair hour in the then fair weather,
And the cave just under, with a voice still so hollow
That it seems to call out to me from forty years ago,
When you were all aglow,
And not the thin ghost that I now frailly follow!

Ignorant of what there is flitting here to see
The waked birds preen and the seals flop lazily,
Soon you will have, Dear, to vanish from me,
For the stars close their shutters and the dawn whitens hazily.
Trust me I mind not, though Life lowers,
The bringing of me here; nay, bring me here again!
I am just the same as when
Our days were a joy, and our paths through flowers.

Pentargan Bay.

7 Hardy's drawing of Beeny Cliff (see p. 74)

8 (above) Beeny Cliff today

9 (below) Hardy's drawing of Salisbury Cathedral ('The Impercipient', see p. 53)

10 Hardy in his study about 1913

In Time of 'The Breaking of Nations'

ONLY a man harrowing clods
  In a slow silent walk
With an old horse that stumbles and nods
  Half asleep as they stalk.

Only thin smoke without flame [5]
  From the heaps of couch-grass;
Yet this will go onward the same
  Though Dynasties pass.

Yonder a maid and her wight
  Come whispering by: [10]
War's annals will cloud into night
  Ere their story die.

1915

'I believe it would be said by people who knew me well that I have a faculty (possibly not uncommon) for burying an emotion in my heart or brain for forty years, and exhuming it at the end of that time as fresh as when interred. For instance, the poem entitled 'The Breaking of Nations' contains a feeling that moved me in 1870, during the Franco-Prussian war, when I chanced to be looking at such an agricultural incident in Cornwall. But I did not write the verses till during the war with Germany of 1914, and onwards. Query: where was that sentiment hiding itself during more than forty years?' (*Life*, p. 378) In this poem Hardy comments on the permanence of such simple things as work and love. Man must cultivate the earth so that he can eat, and he will continue to fall in love. Not even the madness of war can change these basic certainties. This has always been one of Hardy's most popular poems, probably because it states a great truth so simply and effectively.

The Breaking of Nations a reference to Jeremiah 51: 20. 'Thou art my battle axe and weapons of war: for with thee will I break in pieces the nations, and with thee will I destroy kingdoms.'
[6] **couch-grass** a coarse weed-like grass.
[8] **Dynasties** kingdoms.
[9] **wight** man.
Hardy deliberately uses poetic and old-fashioned words (compare 'Yonder' and 'Ere') in order to give a feeling of timelessness.

Ten Years Since

'TIS ten years since
I saw her on the stairs,
Heard her in house-affairs,
And listened to her cares;
And the trees are ten feet taller, [5]
And the sunny spaces smaller
Whose bloomage would enthrall her;
And the piano wires are rustier,
The smell of bindings mustier,
And lofts and lumber dustier [10]
Than when, with casual look
And ear, light note I took
Of what shut like a book
Those ten years since!

November 1922

Nobody Comes

TREE-LEAVES labour up and down,
And through them the fainting light
Succumbs to the crawl of night.
Outside in the road the telegraph wire
To the town from the darkening land [5]
Intones to travellers like a spectral lyre
Swept by a spectral hand.

A car comes up, with lamps full-glare,
That flash upon a tree:
It has nothing to do with me, [10]
And whangs along in a world of its own,
Leaving a blacker air;
And mute by the gate I stand again alone,
And nobody pulls up there.

(*Written on 9 October, 1924*)

Hardy wrote this when he was eighty-four years old. The proportion of descriptive poetry increased as he grew older, but this is not just a descriptive poem. There is in it something of the loneliness of old age.

Night in the Old Home

WHEN the wasting embers redden the chimney-breast,
And Life's bare pathway looms like a desert track to me,
And from hall and parlour the living have gone to their rest,
My perished people who housed them here come back to me.

They come and seat them around in their mouldy places, [5]
Now and then bending towards me a glance of wistfulness,
A strange upbraiding smile upon all their faces,
And in the bearing of each a passive tristfulness.

'Do you uphold me, lingering and languishing here,
A pale late plant of your once strong stock?' I say to them; [10]
'A thinker of crooked thoughts upon Life in the sere,
And on That which consigns men to night after showing the day
to them?'

'– O let be the Wherefore! We fevered our years not thus:
Take of Life what it grants, without question!' they answer me
seemingly.
'Enjoy, suffer, wait: spread the table here freely like us, [15]
And, satisfied, placid, unfretting, watch Time away beamingly!'

Hardy imagines himself back in the old cottage at Higher Bockhampton which he described in 'Domicilium' so many years before. The many ancestors who had lived in the cottage return as ghosts and seem to be reproving him for his inability to accept life without question, as they did. Hardy had no children and sees himself as the last weak descendant of his family. His tone is humble and regretful that he could not be as they were, able to 'watch Time away beamingly!' Perhaps we should be grateful that he was not as they were as the world would have lost many fine poems if he had been.

[7] **upbraiding** reproachful.
[8] **tristfulness** sadness.
[11] **Life in the sere** life past its best, growing old and withered.
[12] **That** The capital letter shows that Hardy means whatever power controls our lives.

A Private Man on Public Men

WHEN my contemporaries were driving
Their coach through Life with strain and striving,
And raking riches into heaps,
And ably pleading in the Courts
With smart rejoinders and retorts, [5]
Or where the Senate nightly keeps
Its vigils, till their fames were fanned
By rumour's tongue throughout the land,
I lived in quiet, screened, unknown,
Pondering upon some stick or stone, [10]
Or news of some rare book or bird
Latterly bought, or seen, or heard,
Not wishing ever to set eyes on
The surging crowd beyond the horizon,
Tasting years of moderate gladness [15]
Mellowed by sundry days of sadness,
Shut from the noise of the world without,
Hearing but dimly its rush and rout,
Unenvying those amid its roar,
Little endowed, not wanting more. [20]

It is a temptation to regard the 'Private Man' as being an autobiographical account of Hardy. Undoubtedly there is something of Hardy in the picture but a writer of poetry is not on oath to tell the factual truth of life. It is certainly not true that he was 'little endowed' or that he lived 'unknown', but in his later years he did live a secluded life and he was not one who at any time would have ruthlessly sought advancement, fame, and wealth.

The Superseded

As newer comers crowd the fore,
    We drop behind.
– We who have laboured long and sore
    Times out of mind,
And keen are yet, must not regret [5]
    To drop behind.

Yet there are some of us who grieve
    To go behind;
Staunch, strenuous souls who scarce believe
    Their fires declined, [10]
And know none spares, remembers, cares
    Who go behind.

'Tis not that we have unforetold
    The drop behind;
We feel the new must oust the old [15]
    In every kind;
But yet we think, must we, must *we*,
    Too, drop behind?

Published in 1901 when Hardy still had nearly thirty years to live but was already feeling that he was being superseded by younger men, the poem is suffused with the sadness of growing older. Hardy was very fond of using a short line after a long one, particularly at the end of a stanza.

[11] **spares** shows mercy to.

He Never Expected Much
(or)
A Consideration
(A reflection) on my eighty-sixth birthday

WELL, World, you have kept faith with me,
        Kept faith with me;
Upon the whole you have proved to be
        Much as you said you were.
Since as a child I used to lie [5]
Upon the leaze and watch the sky,
Never, I own, expected I
        That life would all be fair.

'Twas then you said, and since have said,
        Times since have said, [10]
In that mysterious voice you shed
        From clouds and hills around:
'Many have loved me desperately,
Many with smooth serenity,
While some have shown contempt of me [15]
        Till they dropped underground.

'I do not promise overmuch,
        Child; overmuch;
Just neutral-tinted haps and such,'
        You said to minds like mine. [20]
Wise warning for your credit's sake!
Which I for one failed not to take,
And hence could stem such strain and ache
        As each year might assign.

[6] **leaze** meadow-land.
[19] **haps** happenings.

'I Look into My Glass'

I LOOK into my glass,
And view my wasting skin,
And say, 'Would God it came to pass
My heart had shrunk as thin!'

For then, I, undistrest [5]
By hearts grown cold to me,
Could lonely wait my endless rest
With equanimity.

But Time, to make me grieve,
Part steals, lets part abide; [10]
And shakes this fragile frame at eve
With throbbings of noontide.

[1] **glass** mirror.

Afterwards

WHEN the Present has latched its postern behind my tremulous
stay,
And the May month flaps its glad green leaves like wings,
Delicate-filmed as new-spun silk, will the neighbours say,
'He was a man who used to notice such things'?

If it be in the dusk when, like an eyelid's soundless blink, [5]
The dewfall-hawk comes crossing the shades to alight
Upon the wind-warped upland thorn, a gazer may think,
'To him this must have been a familiar sight.'

If I pass during some nocturnal blackness, mothy and warm,
When the hedgehog travels furtively over the lawn, [10]
One may say, 'He strove that such innocent creatures should
come to no harm,
But he could do little for them; and now he is gone.'

If, when hearing that I have been stilled at last, they stand at the
door,
Watching the full-starred heavens that winter sees,
Will this thought rise on those who will meet my face no more, [15]
'He was one who had an eye for such mysteries'?

And will any say when my bell of quittance is heard in the gloom,
And a crossing breeze cuts a pause in its outrollings,
Till they rise again, as they were a new bell's boom,
'He hears it not now, but used to notice such things'? [20]

[1] **postern** a small or private door.
[17] **bell of quittance** funeral bell.

Incidents and Stories

The Pine Planters
(Marty South's Reverie)

I

WE work here together
    In blast and breeze;
He fills the earth in,
    I hold the trees.

He does not notice [5]
    That what I do
Keeps me from moving
    And chills me through.

He has seen one fairer
    I feel by his eye, [10]
Which skims me as though
    I were not by.

And since she passed here
    He scarce has known
But that the woodland [15]
    Holds him alone.

I have worked here with him
    Since morning shine,
He busy with his thoughts
    And I with mine. [20]

I have helped him so many,
    So many days,
But never win any
    Small word of praise!

Shall I not sigh to him [25]
    That I work on
Glad to be nigh to him
    Though hope is gone?

Nay, though he never
    Knew love like mine, [30]
I'll bear it ever
    And make no sign!

II

From the bundle at hand here
    I take each tree,
And set it to stand, here [35]
    Always to be;
When, in a second,
    As if from fear
Of Life unreckoned
    Beginning here, [40]
It starts a sighing
    Through day and night,
Though while there lying
    'Twas voiceless quite.

It will sigh in the morning, [45]
    Will sigh at noon,
At the winter's warning,
    In wafts of June;
Grieving that never
    Kind Fate decreed [50]
It should for ever
    Remain a seed,
And shun the welter
    Of things without,
Unneeding shelter [55]
    From storm and drought.

Thus, all unknowing
    For whom or what
We set it growing
    In this bleak spot, [60]
It still will grieve here
    Throughout its time,

Unable to leave here,
    Or change its clime:
Or tell the story [65]
    Of us today
When, halt and hoary,
    We pass away.

Marty South is a character in Hardy's novel, *The Woodlanders*. One of the memorable features of the book is her selfless love for Giles Winterborne who loves another. She often helps Giles with his work of planting trees.

[48] **wafts** very small breezes, almost no more than slight movements of air.

Tess's Lament

I WOULD that folk forgot me quite,
    Forgot me quite!
I would that I could shrink from sight,
    And no more see the sun.
Would it were time to say farewell, [5]
To claim my nook, to need my knell,
Time for them all to stand and tell
    Of my day's work as done.

Ah! dairy where I lived so long,
    I lived so long; [10]
Where I would rise up staunch and strong,
    And lie down hopefully.
'Twas there within the chimney-seat
He watched me to the clock's slow beat –
Loved me, and learnt to call me Sweet, [15]
    And whispered words to me.

And now he's gone; and now he's gone; . . .
And now he's gone!
The flowers we potted perhaps are thrown
To rot upon the farm. [20]
And where we had our supper-fire
May now grow nettle, dock, and briar,
And all the place be mould and mire
So cosy once and warm.

And it was I who did it all, [25]
Who did it all;
'Twas I who made the blow to fall
On him who thought no guile.
Well, it is finished – past, and he
Has left me to my misery, [30]
And I must take my Cross on me
For wronging him awhile.

How gay we looked that day we wed,
That day we wed!
'May joy be with ye!' they all said [35]
A-standing by the durn.
I wonder what they say o'us now,
And if they know my lot; and how
She feels who milks my favourite cow,
And takes my place at churn! [40]

It wears me out to think of it,
To think of it;
I cannot bear my fate as writ,
I'd have my life unbe;

Would turn my memory to a blot, [45]
Make every relic of me rot,
My doings be as they were not,
And gone all trace of me!

Tess is the tragic main character of Hardy's novel, *Tess of the d'Urbervilles.* She loves and marries Angel Clare but he leaves her after learning on their wedding-night that she had previously been seduced. A critic said of the poem that it, 'wails in a metre which seems to rock . . . with an infinite haunting sadness'. (Edmund Gosse)

[6] **knell** funeral bell.
[36] **durn** door-post.
[38] **lot** fate.

Boys Then and Now

'MORE than one cuckoo?'
And the little boy
Seemed to lose something
Of his spring joy.

When he'd grown up [5]
He told his son
He'd used to think
There was only one,

Who came each year
With the trees' new trim [10]
On purpose to please
England and him:

And his son – old already
In life and its ways –
Said yawning: 'How foolish [15]
Boys were in those days!'

At the Railway Station, Upway

'THERE is not much that I can do,
  For I've no money that's quite my own!'
  Spoke up the pitying child –
A little boy with a violin
At the station before the train came in, – [5]
'But I can play my fiddle to you,
And a nice one 'tis, and good in tone!'

  The man in the handcuffs smiled;
The constable looked, and he smiled, too,
  As the fiddle began to twang; [10]
And the man in the handcuffs suddenly sang
      With grimful glee:
      'This life so free
      Is the thing for me!'
And the constable smiled, and said no word, [15]
As if unconscious of what he heard;
And so they went on till the train came in –
The convict, and boy with the violin.

Upway (normally spelt 'Upwey') is a village on the river Wey about three miles south of Dorchester. Portland Prison is about three miles further south.

In a Waiting-Room

ON a morning sick as the day of doom
    With the drizzling grey
    Of an English May,
There were few in the railway waiting-room.
About its walls were framed and varnished [5]
Pictures of liners, fly-blown, tarnished.
The table bore a Testament
For travellers' reading, if suchwise bent.

I read it on and on,
And, thronging the Gospel of Saint John, [10]
Were figures – additions, multiplications –
By some one scrawled, with sundry emendations;
Not scoffingly designed,
But with an absent mind, –
Plainly a bagman's counts of cost, [15]
What he had profited, what lost;
And whilst I wondered if there could have been
Any particle of a soul
In that poor man at all,
To cypher rates of wage [20]
Upon that printed page,
There joined in the charmless scene
And stood over me and the scribbled book
(To lend the hour's mean hue
A smear of tragedy too) [25]
A soldier and wife, with haggard look
Subdued to stone by strong endeavour;
And then I heard
From a casual word
They were parting as they believed for ever. [30]

But next there came
Like the eastern flame
Of some high altar, children – a pair –
Who laughed at the fly-blown pictures there.
'Here are the lovely ships that we, [35]
Mother, are by and by going to see!
When we get there it's 'most sure to be fine,
And the band will play, and the sun will shine!'

It rained on the skylight with a din
As we waited and still no train came in; [40]
But the words of the child in the squalid room
Had spread a glory through the gloom.

Note how the drabness of the opening is conveyed in the awkward, stilted

language and rhythm but with the arrival of the children, with their hope and optimism, there is a change in both rhythm and language.
[15] **bagman** commercial traveller.

The Bird-Catcher's Boy

'FATHER, I fear your trade:
  Surely it's wrong!
Little birds limed and made
  Captive life-long.

'Larks bruise and bleed in jail, [5]
  Trying to rise;
Every caged nightingale
  Soon pines and dies.'

'Don't be a dolt, my boy!
  Birds must be caught; [10]
My lot is such employ,
  Yours to be taught.

'Soft shallow stuff as that
  Out from your head!
Just learn your lessons pat, [15]
  Then off to bed.'

Lightless, without a word
  Bedwise he fares;
Groping his way is heard
  Seek the dark stairs [20]

Through the long passage, where
  Hang the caged choirs:
Harp-like his fingers there
  Sweep on the wires.

Next day, at dye of dawn, [25]
Freddy was missed:
Whither the boy had gone
Nobody wist.

That week, the next one, whiled:
No news of him: [30]
Weeks up to months were piled:
Hope dwindled dim.

Yet not a single night
Locked they the door,
Waiting, heart-sick, to sight [35]
Freddy once more.

Hopping there long anon
Still the birds hung:
Like those in Babylon
Captive, they sung. [40]

One wintry Christmastide
Both lay awake;
All cheer within them dried,
Each hour an ache.

Then some one seemed to flit [45]
Soft in below;
'Freddy's come!' Up they sit,
Faces aglow.

Thereat a groping touch
Dragged on the wires [50]
Lightly and softly – much
As they were lyres;

'Just as it used to be
When he came in,
Feeling in darkness the [55]
Stairway to win!'

Waiting a trice or two
  Yet, in the gloom,
Both parents pressed into
  Freddy's old room. [60]

There on the empty bed
  White the moon shone,
As ever since they'd said,
  'Freddy is gone!'

That night at Durdle-Door [65]
  Foundered a hoy,
And the tide washed ashore
  One sailor boy.

21 November, 1912

[3] **limed** Bird-lime was a sticky substance used to catch birds.
[25] **at dye of dawn** i.e. when the colours of dawn dyed the sky.
[29] **whiled** passed.
[39] **Babylon** The captives in Babylon sing in Psalm 137.
[52] **lyres** stringed musical instruments.
[57] **trice** moment.
[65] **Durdle-Door** a rock on the south coast, near Lulworth Cove.
[66] **hoy** a small sailing vessel used to carry passengers and goods.

A Wife Waits

WILL's at the dance in the Club-room below,
    Where the tall liquor-cups foam;
I on the pavement up here by the Bow,
    Wait, wait, to steady him home.

Will and his partner are treading a tune, [5]
    Loving companions they be;
Willy, before we were married in June,
    Said he loved no one but me;

Said he would let his old pleasures all go
    Ever to live with his Dear. [10]
Will's at the dance in the Club-room below,
    Shivering I wait for him here.

[3] **the Bow** the old name for the curved corner by the cross-streets in the middle of Casterbridge (Dorchester). This is Hardy's own note.

The Market-Girl

NOBODY took any notice of her as she stood on the causey kerb
All eager to sell her honey and apples and bunches of garden herb;
And if she had offered to give her wares and herself with them
    too that day,
I doubt if a soul would have cared to take a bargain so choice
    away.

But chancing to trace her sunburnt grace that morning as I
    passed nigh, [5]
I went and I said 'Poor maidy dear! – and will none of the
    people buy?'
And so it began; and soon we knew what the end of it all must
    be,
And I found that though no others had bid, a prize had been
    won by me.

[1] **causey** paved-way.

Throwing a Tree
(New Forest)

THE two executioners stalk along over the knolls,
Bearing two axes with heavy heads shining and wide,
And a long limp two-handled saw toothed for cutting great boles,
And so they approach the proud tree that bears the death-mark on its side.

Jackets doffed they swing axes and chop away just above ground, [5]
And the chips fly about and lie white on the moss and fallen leaves;
Till a broad deep gash in the bark is hewn all the way round,
And one of them tries to hook upward a rope, which at last he achieves.

The saw then begins, till the top of the tall giant shivers:
The shivers are seen to grow greater each cut than before: [10]
They edge out the saw, tug the rope; but the tree only quivers,
And kneeling and sawing again, they step back to try pulling once more.

Then, lastly, the living mast sways, further sways: with a shout
Job and Ike rush aside. Reached the end of its long staying powers
The tree crashes downward: it shakes all its neighbours throughout, [15]
And two hundred years' steady growth has been ended in less than two hours.

Hardy felt that trees had some of the qualities of people, including the ability to suffer. Note how he shows that feeling here by the use of words like 'executioners'.

Throwing cutting down.
New Forest a well-known forest in Hampshire.
[1] **knolls** small hills or mounds.
[3] **boles** tree-trunks.

The Old Workman

'WHY are you so bent down before your time,
Old mason? Many have not left their prime
So far behind at your age, and can still
        Stand full upright at will.'

He pointed to the mansion-front hard by, [5]
And to the stones of the quoin against the sky;
'Those upper blocks,' he said, 'that there you see,
        It was that ruined me.'

There stood in the air up to the parapet
Crowning the corner height, the stones as set [10]
By him – ashlar whereon the gales might drum
        For centuries to come.

'I carried them up,' he said, 'by a ladder there;
The last was as big a load as I could bear;
But on I heaved; and something in my back [15]
        Moved, as 'twere with a crack.

'So I got crookt. I never lost that sprain;
And those who live there, walled from wind and rain
By freestone that I lifted, do not know
        That my life's ache came so. [20]

'They don't know me, or even know my name,
But good I think it, somehow, all the same
To have kept 'em safe from harm, and right and tight,
        Though it has broke me quite.

'Yes; that I fixed it firm up there I am proud, [25]
Facing the hail and snow and sun and cloud,
And to stand storms for ages, beating round
        When I lie underground.'

Hardy's sympathy for and understanding of the working-man is a distinctive feature of his novels and poems. There is nothing sentimental about this. He appreciates the hardness and the strains of the working-man's life, but he is also aware of his pride in his work and his toughness and courage.

[2] **mason** i.e. stone-mason or builder.
[6] **quoin** the external corner of a building.
[11] **ashlar** stone used in facing a wall.
[19] **freestone** building stone.

The Choirmaster's Burial

HE often would ask us
That, when he died,
After playing so many
To their last rest,
If out of us any [5]
Should here abide,
And it would not task us,
We would with our lutes
Play over him
By his grave-brim [10]
The psalm he liked best –
The one whose sense suits
'Mount Ephraim.' –
And perhaps we should seem
To him, in Death's dream, [15]
Like the seraphim.

As soon as I knew
That his spirit was gone
I thought this his due,
And spoke thereupon. [20]

'I think,' said the vicar,
'A read service quicker
Than viols out-of-doors
In these frosts and hoars.

That old-fashioned way [25]
Requires a fine day,
And it seems to me
It had better not be.'

Hence, that afternoon,
Though never knew he [30]
That his wish could not be,
To get through it faster
They buried the master
Without any tune.

But 'twas said that, when [35]
At the dead of next night
The vicar looked out,
There struck on his ken
Thronged roundabout,
Where the frost was greying [40]
The headstoned grass,
A band all in white
Like the saints in church-glass,
Singing and playing
The ancient stave [45]
By the choirmaster's grave.

Such the tenor man told
When he had grown old.

At the time of Hardy's grandfather and father, the first forty years of the nineteenth century, church choirs often included instrumentalists, but these were gradually ousted by the organ. Hardy's novel, *Under the Greenwood Tree,* is in part about the dying of this old tradition of having instrumentalists.

[11] **psalm . . . best** 'This psalm is "The Lord God Jehovah reigns".' (J. O. Bailey).
[13] **Mount Ephraim** a popular psalm tune.
[16] **seraphim** the highest order of angels.
[23] **viols** stringed instruments played with a bow.
[24] **hoars** another name for frosts.
[38] **ken** sight.
[45] **stave** tune.

The Sacrilege

A Ballad-tragedy
(About 182–)

PART I

'I HAVE a Love I love too well
Where Dunkery frowns on Exon Moor;
I have a Love I love too well,
To whom, ere she was mine,
"Such is my love for you," I said, [5]
"That you shall have to hood your head
A silken kerchief crimson-red,
Wove finest of the fine."

'And since this Love, for one mad moon,
On Exon Wild by Dunkery Tor, [10]
Since this my Love for one mad moon
Did clasp me as her king,
I snatched a silk-piece red and rare
From off a stall at Priddy Fair,
For handkerchief to hood her hair [15]
When we went gallanting.

'Full soon the four weeks neared their end
Where Dunkery frowns on Exon Moor;
And when the four weeks neared their end,
And their swift sweets outwore, [20]
I said, "What shall I do to own
Those beauties bright as tulips blown,
And keep you here with me alone
As mine for evermore?"

'And as she drowsed within my van [25]
On Exon Wild by Dunkery Tor –
And as she drowsed within my van,
And dawning turned to day,

She heavily raised her sloe-black eyes
And murmured back in softest wise, [30]
"One more thing, and the charms you prize
        Are yours henceforth for aye.

' "And swear I will I'll never go
While Dunkery frowns on Exon Moor
To meet the Cornish Wrestler Joe [35]
        For dance and dallyings.
If you'll to yon cathedral shrine,
And finger from the chest divine
Treasure to buy me ear-drops fine,
        And richly jewelled rings." [40]

'I said: "I am one who has gathered gear
From Marlbury Downs to Dunkery Tor,
Who has gathered gear for many a year
        From mansion, mart and fair;
But at God's house I've stayed my hand, [45]
Hearing within me some command –
Curbed by a law not of the land
        From doing damage there!"

'Whereat she pouts, this Love of mine,
As Dunkery pouts to Exon Moor, [50]
And still she pouts, this Love of mine,
        So cityward I go.
But ere I start to do the thing,
And speed my soul's imperilling
For one who is my ravishing [55]
        And all the joy I know,

'I come to lay this charge on thee –
On Exon Wild by Dunkery Tor –
I come to lay this charge on thee
        With solemn speech and sign: [60]

Should things go ill, and my life pay
For botchery in this rash assay,
You are to take hers likewise – yea,
The month the law takes mine.

'For should my rival, Wrestler Joe, [65]
Where Dunkery frowns on Exon Moor –
My reckless rival, Wrestler Joe,
My Love's bedwinner be,
My tortured spirit would not rest,
But wander weary and distrest [70]
Throughout the world in wild protest:
The thought nigh maddens me!

PART II

Thus did he speak – this brother of mine –
On Exon Wild by Dunkery Tor,
Born at my birth of mother of mine, [75]
And forthwith went his way
To dare the deed some coming night . . .
I kept the watch with shaking sight,
The moon at moments breaking bright,
At others glooming grey. [80]

For three full days I heard no sound
Where Dunkery frowns on Exon Moor,
I heard no sound at all around
Whether his fay prevailed,
Or one more foul the master were, [85]
Till some afoot did tidings bear
How that, for all his practised care,
He had been caught and jailed.

They had heard a crash when twelve had chimed
By Mendip east of Dunkery Tor, [90]
When twelve had chimed and moonlight climbed;
They watched, and he was tracked

By arch and aisle and saint and knight
Of sculptured stonework sheeted white
In the cathedral's ghostly light, [95]
        And captured in the act.

Yes; for this Love he loved too well
Where Dunkery sights the Severn shore,
All for this Love he loved too well
        He burst the holy bars, [100]
Seized golden vessels from the chest
To buy her ornaments of the best,
At her ill-witchery's request
        And lure of eyes like stars. . . .

When blustering March confused the sky [105]
In Toneborough Town by Exon Moor,
When blustering March confused the sky
        They stretched him; and he died.
Down in the crowd where I, to see
The end of him, stood silently, [110]
With a set face he lipped to me –
        'Remember.' 'Ay!' I cried.

By night and day I shadowed her
From Toneborough Deane to Dunkery Tor,
I shadowed her asleep, astir, [115]
        And yet I could not bear –
Till Wrestler Joe anon began
To figure as her chosen man,
And took her to his shining van –
        To doom a form so fair! [120]

He made it handsome for her sake –
And Dunkery smiled to Exon Moor –
He made it handsome for her sake,
        Painting it out and in;

And on the door of apple-green [125]
A bright brass knocker soon was seen,
And window-curtains white and clean
For her to sit within.

And all could see she clave to him
As cleaves a cloud to Dunkery Tor, [130]
Yea, all could see she clave to him,
And every day I said,
'A pity it seems to part those two
That hourly grow to love more true:
Yet she's the wanton woman who [135]
Sent one to swing till dead!'

That blew to blazing all my hate,
While Dunkery frowned on Exon Moor,
And when the river swelled, her fate
Came to her pitilessly. . . . [140]
I dogged her, crying: 'Across that plank
They use as bridge to reach yon bank
A coat and hat lie limp and dank;
Your goodman's, can they be?'

She paled, and went, I close behind – [145]
And Exon frowned to Dunkery Tor,
She went, and I came up behind
And tipped the plank that bore
Her, fleetly flitting across to eye
What such might bode. She slid awry; [150]
And from the current came a cry,
A gurgle; and no more.

How that befell no mortal knew
From Marlbury Downs to Exon Moor;
No mortal knew that deed undue [155]
But he who schemed the crime,

Which night still covers. . . . But in dream
Those ropes of hair upon the stream
He sees, and he will hear that scream
Until his judgement-time. [160]

Sacrilege breaking into a holy place and stealing.
[2] **Dunkery Tor** Dunkery Beacon is the highest hill on Exmoor.
Exon Moor Exmoor.
[7] **kerchief** a cloth covering for the head.
[14] **Priddy** a village in the Mendip Hills.
[41] **gear** stolen property.
[42] **Marlbury** Marlborough.
[84] **fay** good fairy.
[106] **Toneborough Town** Taunton.
[108] **stretched** hanged.

A Trampwoman's Tragedy
(182–)

FROM Wynyard's Gap the livelong day,
The livelong day,
We beat afoot the northward way
We had travelled times before.
The sun-blaze burning on our backs, [5]
Our shoulders sticking to our packs,
By fosseway, fields, and turnpike tracks
We skirted sad Sedge-Moor.

Full twenty miles we jaunted on,
We jaunted on, – [10]
My fancy-man, and jeering John,
And Mother Lee, and I.
And, as the sun drew down to west,
We climbed the toilsome Poldon crest,
And saw, of landskip sights the best, [15]
The inn that beamed thereby.

For months we had padded side by side,
        Ay, side by side
Through the Great Forest, Blackmoor wide,
    And where the Parret ran. [20]
We'd faced the gusts on Mendip ridge,
Had crossed the Yeo unhelped by bridge,
Been stung by every Marshwood midge,
    I and my fancy-man.

Lone inns we loved, my man and I, [25]
        My man and I;
'King's Stag', 'Windwhistle' high and dry,
    'The Horse' on Hintock Green,
The cosy house at Wynyard's Gap,
'The Hut' renowned on Bredy Knap, [30]
And many another wayside tap
    Where folk might sit unseen.

Now as we trudged – O deadly day,
        O deadly day! –
I teased my fancy-man in play [35]
    And wanton idleness.
I walked alongside jeering John,
I laid his hand my waist upon;
I would not bend my glances on
    My lover's dark distress. [40]

Thus Poldon top at last we won,
        At last we won,
And gained the inn at sink of sun
    Far-famed as 'Marshal's Elm'.
Beneath us figured tor and lea, [45]
From Mendip to the western sea –
I doubt if finer sight there be
    Within this royal realm.

Inside the settle all a-row –
All four a-row [50]
We sat, I next to John, to show
That he had wooed and won.
And then he took me on his knee,
And swore it was his turn to be
My favoured mate, and Mother Lee [55]
Passed to my former one.

Then in a voice I had never heard,
I had never heard,
My only Love to me: 'One word,
My lady, if you please! [60]
Whose is the child you are like to bear? –
His? After all my months o' care?'
God knows 'twas not! But, O despair!
I nodded – still to tease.

Then up he sprung, and with his knife – [65]
And with his knife
He let out jeering Johnny's life,
Yes; there, at set of sun.
The slant ray through the window nigh
Gilded John's blood and glazing eye, [70]
Ere scarcely Mother Lee and I
Knew that the deed was done.

The taverns tell the gloomy tale,
The gloomy tale,
How that at Ivel-chester jail [75]
My Love, my sweetheart swung;
Though stained till now by no misdeed
Save one horse ta'en in time o' need;
(Blue Jimmy stole right many a steed
Ere his last fling he flung.) [80]

Thereaft I walked the world alone,
Alone, alone!
On his death-day I gave my groan
And dropt his dead-born child.
'Twas nigh the jail, beneath a tree, [85]
None tending me; for Mother Lee
Had died at Glaston, leaving me
Unfriended on the wild.

And in the night as I lay weak,
As I lay weak, [90]
The leaves a-falling on my cheek,
The red moon low declined –
The ghost of him I'd die to kiss
Rose up and said: 'Ah, tell me this!
Was the child mine, or was it his? [95]
Speak, that I rest may find!'

O doubt not but I told him then,
I told him then,
That I had kept me from all men
Since we joined lips and swore. [100]
Whereat he smiled, and thinned away
As the wind stirred to call up day . . .
– 'Tis past! And here alone I stray
Haunting the Western Moor.

April 1902

'Hardy considered this, upon the whole, his most successful poem.' (*Life*, p. 312) It was offered first to the *Cornhill* magazine but rejected on the ground that it was a poem that could not possibly be published in a family periodical.

[1] **Wynyard's Gap** a few miles north of Beaminster in Dorset.
[7] **fosseway** Roman road.
[8] **Sedge-Moor** Sedgemoor, near Bridgwater in Somerset, the scene of a bloody battle in 1685.
[11] **fancy-man** lover.
[14] **Poldon crest** Poldon Hill in Somerset.

[19] **the Great Forest** the New Forest in Hampshire.

[20] **Parret** a river which rises in Dorset and flows into the Bristol Channel.

[22] **Yeo** a river which rises near Sherborne and flows into the Parret at Langport.

[23] **Marshwood** the vale of Marshwood is in southwest Dorset.

[27] **'Windwhistle'** Hardy provides the following note: 'The highness and dryness of Windwhistle Inn was impressed upon the writer two or three years ago, when, after climbing on a hot afternoon to the beautiful spot near which it stands and entering the inn for tea, he was informed by the landlady that none could be had, unless he would fetch water from a valley half a mile off, the house containing not a drop, owing to its situation. However, a tantalising row of full barrels behind her back testified to a wetness of a certain sort, which was not at that time desired.'

[30] **Bredy Knap** Long Bredy in Dorset.

[31] **tap** inn.

[44] **'Marshal's Elm'** Hardy provides the following note: 'Marshal's Elm, so picturesquely situated, is no longer an inn, though the house, or part of it, still remains. It used to exhibit a fine old swinging sign.'

[45] **tor and lea** hill and meadow.

[49] **settle** a long high-backed bench.

[75] **Ivel-chester** Ilchester on the river Yeo.

[79] **Blue Jimmy** Hardy provides the following note: 'Blue Jimmy was a notorious horse-stealer of Wessex in those days, who appropriated more than a hundred horses before he was caught, among others one belonging to a neighbour of the writer's grandfather. He was hanged at the now demolished Ivel-chester or Ilchester jail above mentioned – that building formerly of so many sinister associations in the minds of the local peasantry, and the continual haunt of fever, which at last led to its condemnation. Its site is now an innocent-looking green meadow.'

[84] **dropt** gave birth to.

[87] **Glaston** Glastonbury.

[104] **Western Moor** probably Exmoor.

A Sunday Morning Tragedy
(about 186–)

I BORE a daughter flower-fair,
In Pydel Vale, alas for me;
I joyed to mother one so rare,
But dead and gone I now would be.

Men looked and loved her as she grew, [5]
And she was won, alas for me;
She told me nothing, but I knew,
And saw that sorrow was to be.

I knew that one had made her thrall,
A thrall to him, alas for me; [10]
And then, at last, she told me all,
And wondered what her end would be.

She owned that she had loved too well,
Had loved too well, unhappy she,
And bore a secret time would tell, [15]
Though in her shroud she'd sooner be.

I plodded to her sweetheart's door
In Pydel Vale, alas for me:
I pleaded with him, pleaded sore,
To save her from her misery. [20]

He frowned, and swore he could not wed,
Seven times he swore it could not be;
'Poverty's worse than shame,' he said,
Till all my hope went out of me.

'I've packed my traps to sail the main' – [25]
Roughly he spake, alas did he –
'Wessex beholds me not again,
'Tis worse than any jail would be!'

– There was a shepherd whom I knew,
A subtle man, alas for me: [30]
I sought him all the pastures through,
Though better I had ceased to be.

I traced him by his lantern light,
And gave him hint, alas for me,
Of how she found her in the plight [35]
That is so scorned in Christendie.

'Is there an herb. . . . ?' I asked. 'Or none?'
Yes, thus I asked him desperately.
'– There is,' he said; 'a certain one. . . .'
Would he had sworn that none knew he! [40]

'Tomorrow I will walk your way,'
He hinted low, alas for me. –
Fieldwards I gazed throughout next day;
Now fields I never more would see!

The sunset-shine, as curfew strook, [45]
As curfew strook beyond the lea,
Lit his white smock and gleaming crook,
While slowly he drew near to me.

He pulled from underneath his smock
The herb I sought, my curse to be – [50]
'At times I use it in my flock,'
He said, and hope waxed strong in me.

''Tis meant to balk ill-motherings' –
(Ill-motherings! Why should they be?) –
'If not, would God have sent such things?' [55]
So spoke the shepherd unto me.

That night I watched the poppling brew,
With bended back and hand on knee:
I stirred it till the dawnlight grew,
And the wind whiffled wailfully. [60]

'This scandal shall be slain,' said I,
'That lours upon her innocency:
I'll give all whispering tongues the lie'; –
But worse than whispers was to be.

'Here's physic for untimely fruit,' [65]
I said to her, alas for me,
Early that morn in fond salute;
And in my grave I now would be.

– Next Sunday came, with sweet church chimes
In Pydel Vale, alas for me: [70]
I went into her room betimes;
No more may such a Sunday be!

'Mother, instead of rescue nigh,'
She faintly breathed, alas for me,
'I feel as I were like to die, [75]
And underground soon, soon should be.'

From church that noon the people walked
In twos and threes, alas for me,
Showed their new raiment – smiled and talked,
Though sackcloth-clad I longed to be. [80]

Came to my door her lover's friends,
And cheerly cried, alas for me,
'Right glad are we he makes amends,
For never a sweeter bride can be.'

My mouth dried, as 'twere scorched within, [85]
Dried at their words, alas for me:
More and more neighbours crowded in,
(O why should mothers ever be!)

'Ha-ha! Such well-kept news!' laughed they,
Yes – so they laughed, alas for me. [90]
'Whose banns were called in church today?' –
Christ, how I wished my soul could flee!

'Where is she? O the stealthy miss,'
Still bantered they, alas for me,
'To keep a wedding close as this . . .' [95]
Ay, Fortune worked thus wantonly!

'But you are pale – you did not know?'
They archly asked, alas for me,
I stammered, 'Yes – some days – ago,'
While coffined clay I wished to be. [100]

''Twas done to please her, we surmise?'
(They spoke quite lightly in their glee)
'Done by him as a fond surprise?'
I thought their words would madden me.

Her lover entered. 'Where's my bird? – [105]
My bird – my flower – my picotee?
First time of asking, soon the third!'
Ah, in my grave I well may be.

To me he whispered: 'Since your call –'
So spoke he then, alas for me – [110]
'I've felt for her, and righted all.'
– I think of it to agony.

'She's faint today – tired – nothing more –'
Thus did I lie, alas for me. . . .
I called her at her chamber door [115]
As one who scarce had strength to be.

No voice replied. I went within –
O women! scourged the worst are we. . . .
I shrieked. The others hastened in
And saw the stroke there dealt on me. [120]

There she lay – silent, breathless, dead,
Stone dead she lay – wronged, sinless she! –
Ghost-white the cheeks once rosy-red:
Death had took her. Death took not me.

I kissed her colding face and hair, [125]
I kissed her corpse – the bride to be! –
My punishment I cannot bear,
But pray God *not* to pity me.

January 1904

Because of its subject – death through an abortion – this poem was rejected by two of the leading magazines before Hardy managed to get it published in 1908. He defended it by saying, '. . . nobody can say that the treatment is other than moral, and the crime is one of growing prevalence, as you probably know, and the false shame which leads to it is produced by the hypocrisy of the age.' Hardy's compassion, his sense of irony in that the girl's death was unnecessary, and his questioning of the social conventions of his time, are all very apparent. His immense technical competence is shown in the use of the same rhyme for alternate lines throughout the poem, and in the way in which the refrain-like 'alas for me' makes us aware of the mother's suffering.

[2] **Pydel Vale** the valley of the river Piddle which runs a few miles from Dorchester in Dorset.
[9] **made her thrall** i.e. he had captured her love.
[16] **shroud** the dress worn by the dead.
[25] **traps** personal belongings.
[30] **subtle** crafty, clever.
[36] **Christendie** Christendom.
[45] **curfew** a bell rung in the evening.
strook struck.
[53] **ill-motherings** i.e. becoming a mother without being married.
[65] **Here's . . . fruit** 'Here's medicine to end an untimely pregnancy.'
[71] **betimes** in good time.
[91] **banns** a proclamation of the intention to get married.
[96] **wantonly** cruelly, unfeelingly.
[106] **picotee** a variety of carnation.
[107] **soon the third** The banns are asked on three successive Sundays.

The Lost Pyx
A Mediaeval Legend

SOME say the spot is banned: that the pillar Cross-and-Hand
 Attests to a deed of hell;
But of else than of bale is the mystic tale
 That ancient Vale-folk tell.

Ere Cernel's Abbey ceased hereabout there dwelt a priest, [5]
 (In later life sub-prior
Of the brotherhood there, whose bones are now bare
 In the field that was Cernel choir).

One night in his cell at the foot of yon dell
The priest heard a frequent cry: [10]
'Go, father, in haste to the cot on the waste,
And shrive a man waiting to die.'

Said the priest in a shout to the caller without,
'The night howls, the tree-trunks bow;
One may barely by day track so rugged a way, [15]
And can I then do so now?'

No further word from the dark was heard,
And the priest moved never a limb;
And he slept and dreamed; till a Visage seemed
To frown from Heaven at him. [20]

In a sweat he arose; and the storm shrieked shrill,
And smote as in savage joy;
While High-Stoy trees twanged to Bubb-Down Hill,
And Bubb-Down to High-Stoy.

There seemed not a holy thing in hail, [25]
Nor shape of light or love,
From the Abbey north of Blackmore Vale
To the Abbey south thereof.

Yet he plodded thence through the dark immense,
And with many a stumbling stride [30]
Through copse and briar climbed nigh and nigher
To the cot and the sick man's side.

When he would have unslung the Vessels uphung
To his arm in the steep ascent,
He made loud moan: the Pyx was gone [35]
Of the Blessed Sacrament.

Then in dolorous dread he beat his head:
'No earthly prize or pelf
Is the thing I've lost in tempest tossed,
But the Body of Christ Himself!' [40]

He thought of the Visage his dream revealed,
  And turned towards whence he came,
Hands groping the ground along foot-track and field,
  And head in a heat of shame.

Till here on the hill, betwixt vill and vill, [45]
  He noted a clear straight ray
Stretching down from the sky to a spot hard by,
  Which shone with the light of day.

And gathered around the illumined ground
  Were common beasts and rare, [50]
All kneeling at gaze, and in pause profound
  Attent on an object there.

'Twas the Pyx, unharmed 'mid the circling rows
  Of Blackmore's hairy throng,
Whereof were oxen, sheep, and does, [55]
  And hares from the brakes among;

And badgers grey, and conies keen,
  And squirrels of the tree,
And many a member seldom seen
  Of Nature's family. [60]

The ireful winds that scoured and swept
  Through coppice, clump, and dell,
Within that holy circle slept
  Calm as in hermit's cell.

Then the priest bent likewise to the sod [65]
  And thanked the Lord of Love,
And Blessed Mary, Mother of God,
  And all the saints above.

And turning straight with his priceless freight,
  He reached the dying one, [70]
Whose passing sprite had been stayed for the rite
  Without which bliss hath none.

And when by grace the priest won place,
    And served the Abbey well,
He reared this stone to mark where shone [75]
    That midnight miracle.

Legend Hardy provides the note: 'On a lonely table-land above the Vale of Blackmore, between High-Stoy and Bubb-Down hills, and commanding in clear weather views that extend from the English to the Bristol Channel, stands a pillar, apparently medieval, called Cross-and-Hand, or Christ-in-Hand. One tradition of its origin is mentioned in *Tess of the d'Urbervilles*; another, more detailed, preserves the story here given.' The stone is on Batcombe Hill, about eight miles from Dorchester.

[1] **banned** cursed.
[3] **else . . . bale** other than of evil.
[5] **Cernel's Abbey** the Abbey at Cerne Abbas in Dorset.
[6] **sub-prior** A prior was the head of a priory of monks.
[11] **cot** cottage.
[12] **shrive** hear the confession of the man and forgive him his sins.
[33] **Vessels** i.e. the sacred vessels containing the Holy Sacrament (the consecrated bread).
[35] **Pyx** the vessel in which the consecrated bread – for Roman Catholics this is the body of Christ – is kept.
[38] **pelf** money, goods.
[45] **vill** village.
[52] **Attent** intent.
[56] **brakes** clumps of bushes.
[57] **conies** rabbits.
[61] **ireful** angry.
[71] **passing sprite** dying spirit.

the tenhiqu in the rithyme
give a feeling of age

Descriptive and Animal Poems

Weathers

THIS is the weather the cuckoo likes,
    And so do I;
When showers betumble the chestnut spikes,
    And nestlings fly:
And the little brown nightingale bills his best, [5]
And they sit outside at 'The Travellers' Rest',
And maids come forth sprig-muslin drest,
And citizens dream of the south and west,
    And so do I.

This is the weather the shepherd shuns, [10]
    And so do I;
When beeches drip in browns and duns,
    And thresh, and ply;
And hill-hid tides throb, throe on throe,
And meadow rivulets overflow, [15]
And drops on gate-bars hang in a row,
And rooks in families homeward go,
    And so do I.

Hardy liked to build his poems around a contrast – here that between good and bad weather.

[5] **bills** sings.
[7] **sprig-muslin drest** wearing dresses of cotton decorated with a design of sprays or twigs of plants.
[8] **south and west** the south and west of England where people go for their holidays.
[12] **duns** dull greyish-brown colours.
[13] **thresh, and ply** toss and bend.
[14] **hill-hid . . . throe** the sea on the other side of the hill beats in spasms (on the beach).

Before and After Summer

LOOKING forward to the spring
One puts up with anything.
On this February day
Though the winds leap down the street
Wintry scourgings seem but play, [5]
And these later shafts of sleet
– Sharper pointed than the first –
And these later snows – the worst –
Are as a half-transparent blind
Riddled by rays from sun behind. [10]

Shadows of the October pine
Reach into this room of mine:
On the pine there swings a bird;
He is shadowed with the tree.
Mutely perched he bills no word; [15]
Blank as I am even is he.
For those happy suns are past,
Fore-discerned in winter last.
When went by their pleasure, then?
I, alas, perceived not when. [20]

[15] **bills** sings.
[18] **Fore-discerned . . . last** which last winter we saw ahead of us.

A Backward Spring

THE trees are afraid to put forth buds,
And there is timidity in the grass;
The plots lie grey where gouged by spuds,
    And whether next week will pass
Free of sly sour winds is the fret of each bush [5]
    Of barberry waiting to bloom.

Yet the snowdrop's face betrays no gloom,
And the primrose pants in its heedless push,
Though the myrtle asks if it's worth the fight
    This year with frost and rime [10]
    To venture one more time
On delicate leaves and buttons of white
From the selfsame bough as at last year's prime,
And never to ruminate on or remember
What happened to it in mid-December. [15]

April 1917

[3] **spuds** digging and weeding implements.
[6] **barberry** a shrub with small yellow flowers followed by red berries.
[10] **rime** freezing mist.
[14] **ruminate** ponder.

'If It's Ever Spring Again'
(Song)

If it's ever spring again,
    Spring again,
I shall go where went I when
Down the moor-cock splashed, and hen,
Seeing me not, amid their flounder, [5]
Standing with my arm around her;
If it's ever spring again,
    Spring again,
I shall go where went I then.

If it's ever summer-time, [10]
    Summer-time,
With the hay crop at the prime,
And the cuckoos – two – in rhyme,
As they used to be, or seemed to,
We shall do as long we've dreamed to, [15]
If it's ever summer-time,
    Summer-time,
With the hay, and bees achime.

An Unkindly May

A SHEPHERD stands by a gate in a white smock-frock:
He holds the gate ajar, intently counting his flock.

The sour spring wind is blurting boisterous-wise,
And bears on it dirty clouds across the skies;
Plantation timbers creak like rusty cranes, [5]
And pigeons and rooks, dishevelled by late rains,
Are like gaunt vultures, sodden and unkempt,
And song-birds do not end what they attempt:
The buds have tried to open, but quite failing
Have pinched themselves together in their quailing. [10]
The sun frowns whitely in eye-trying flaps
Through passing cloud-holes, mimicking audible taps.
'Nature, you're not commendable today!'
I think. 'Better tomorrow!' she seems to say.

That shepherd still stands in that white smock-frock, [15]
Unnoting all things save the counting his flock.

[1] **smock-frock** a loose-fitting outer-garment which used to be worn by farm-workers.
[10] **quailing** fear or failure.

An August Midnight

A SHADED lamp and a waving blind,
And the beat of a clock from a distant floor:
On this scene enter – winged, horned, and spined –
A longlegs, a moth, and a dumbledore;
While 'mid my page there idly stands [5]
A sleepy fly, that rubs its hands . . .

Thus meet we five, in this still place,
At this point of time, at this point in space.
– My guests besmear my new-penned line,
Or bang at the lamp and fall supine. [10]
'God's humblest, they!' I muse. Yet why?
They know Earth-secrets that know not I.

Max Gate, 1899

'A part of Hardy's unique quality resides in his response to . . . the clock, the darkness, the point in time, the point in space; in his power to invest poetry with this presentness of the present moment.' (Douglas Brown)

[4] **dumbledore** This could be a bumble-bee but is more likely to be a flying beetle called a cockchafer.
[10] **supine** on their backs.

Shortening Days at the Homestead

THE first fire since the summer is lit, and is smoking into the room:
The sun-rays thread it through, like woof-lines in a loom.
Sparrows spurt from the hedge, whom misgivings appal
That winter did not leave last year for ever, after all.
Like shock-headed urchins, spiny-haired, [5]
Stand pollard willows, their twigs just bared.

Who is this coming with pondering pace,
Black and ruddy, with white embossed,
His eyes being black, and ruddy his face
And the marge of his hair like morning frost? [10]
It's the cider-maker,
And appletree-shaker,
And behind him on wheels, in readiness,
His mill, and tubs, and vat, and press.

[2] **woof-lines** In weaving on a loom the woof-lines are the lines of thread which are woven into the warp or lengthwise threads.
[6] **pollard willows** willow trees which have been cut back to the trunk to promote a thick close growth of young branches.
[10] **marge** margin.
[14] **vat** a vessel for storing the cider.

The Later Autumn

GONE are the lovers, under the bush
Stretched at their ease;
Gone the bees,
Tangling themselves in your hair as they rush
On the line of your track, [5]
Leg-laden, back
With a dip to their hive
In a prepossessed dive.

Toadsmeat is mangy, frosted, and sere;
Apples in grass [10]
Crunch as we pass,
And rot ere the men who make cyder appear.
Couch-fires abound
On fallows around,
And shades far extend [15]
Like lives soon to end.

Spinning leaves join the remains shrunk and brown
Of last year's display
That lie wasting away,
On whose corpses they earlier as scorners gazed down [20]
From their aery green height:
Now in the same plight
They huddle; while yon
A robin looks on.

[8] **prepossessed** confident.
[9] **Toadsmeat** toadstools.
mangy scabby.
sere withered.
[13] **Couch-fires** fires lit to burn up couch (i.e. coarse) grass.
[14] **fallows** ploughed but uncultivated land.

personification

Last Week in October

THE trees are undressing, and fling in many places –
On the grey road, the roof, the window-sill –
Their radiant robes and ribbons and yellow laces;
A leaf each second so is flung at will,
Here, there, another and another, still and still. [5]

A spider's web has caught one while downcoming,
That stays there dangling when the rest pass on;
Like a suspended criminal hangs he, mumming
In golden garb, while one yet green, high yon,
Trembles, as fearing such a fate for himself anon. [10]

[8] **mumming** acting (as in a play).
[9] **yon** yonder.
[10] **anon** soon.

The Last Chrysanthemum

WHY should this flower delay so long
To show its tremulous plumes?
Now is the time of plaintive robin-song,
When flowers are in their tombs.

Through the slow summer, when the sun [5]
Called to each frond and whorl
That all he could for flowers was being done,
Why did it not uncurl?

It must have felt that fervid call
Although it took no heed, [10]
Waking but now, when leaves like corpses fall,
And saps all retrocede.

Too late its beauty, lonely thing,
The season's shine is spent,
Nothing remains for it but shivering [15]
In tempests turbulent.

Had it a reason for delay,
Dreaming in witlessness
That for a bloom so delicately gay
Winter would stay its stress? [20]

– I talk as if the thing were born
With sense to work its mind;
Yet it is but one mask of many worn
By the Great Face behind.

[6] **frond** leaf.
whorl a ring of leaves or flowers round a stem.
[12] **retrocede** recede.

Winter Night in Woodland
(Old Time)

THE bark of a fox rings, sonorous and long:—
Three barks, and then silentness; 'wong, wong, wong!'
In quality horn-like, yet melancholy,
As from teachings of years; for an old one is he.
The hand of all men is against him, he knows; and yet, why? [5]
That he knows not, – will never know, down to his death-halloo cry.

With clap-nets and lanterns off start the bird-baiters,
In trim to make raids on the roosts in the copse,
Where they beat the boughs artfully, while their awaiters
Grow heavy at home over divers warm drops. [10]
The poachers, with swingels, and matches of brimstone, outcreep
To steal upon pheasants and drowse them a-perch and asleep.

Out there, on the verge, where a path wavers through,
Dark figures, filed singly, thrid quickly the view,
Yet heavily laden: land-carriers are they [15]
In the hire of the smugglers from some nearest bay.
Each bears his two 'tubs', slung across, one in front, one behind,
To a further snug hiding, which none but themselves are to find.

And then, when the night has turned twelve the air brings
From dim distance, a rhythm of voices and strings: [20]
'Tis the quire, just afoot on their long yearly rounds,
To rouse by worn carols each house in their bounds;
Robert Penny, the Dewys, Mail, Voss, and the rest; till anon
Tired and thirsty, but cheerful, they home to their beds in the dawn.

[7] **clap-nets** nets made to clap together suddenly by pulling a string.
[8] **in trim** well-equipped.
[9] **their awaiters** people waiting for them.
[10] **divers warm drops** various warm drinks.
[11] **swingels** cudgels.
matches of brimstone sulphur matches.
[12] **drowse** stupefy.
[14] **filed singly** in single file.
thrid thread.
[17] **'tubs'** containing strong drink, probably brandy.

Ice on the Highway

SEVEN buxom women abreast, and arm in arm,
Trudge down the hill, tip-toed,
And breathing warm;
They must perforce trudge thus, to keep upright
On the glassy ice-bound road, [5]

And they must get to market whether or no,
Provisions running low
With the nearing Saturday night,
While the lumbering van wherein they mostly ride
Can nowise go: [10]
Yet loud their laughter as they stagger and slide!

Yell'ham Hill

Snow in the Suburbs

EVERY branch big with it,
Bent every twig with it;
Every fork like a white web-foot;
Every street and pavement mute:
Some flakes have lost their way, and grope back upward, when [5]
Meeting those meandering down they turn and descend again.
The palings are glued together like a wall,
And there is no waft of wind with the fleecy fall.

A sparrow enters the tree,
Whereon immediately [10]
A snow-lump thrice his own slight size
Descends on him and showers his head and eyes,
And overturns him,
And near inurns him,
And lights on a nether twig, when its brush [15]
Starts off a volley of other lodging lumps with a rush.

The steps are a blanched slope,
Up which, with feeble hope,
A black cat comes, wide-eyed and thin;
And we take him in. [20]

[14] **inurns** entombs.

To a Tree in London
(Clement's Inn)

  Here you stay
  Night and day,
Never, never going away!

  Do you ache
  When we take   [5]
Holiday for our health's sake?

  Wish for feet
  When the heat
Scalds you in the brick-built street,

  That you might   [10]
  Climb the height
Where your ancestry saw light,

  Find a brook
  In some nook
There to purge your swarthy look?   [15]

  No. You read
  Trees to need
Smoke like earth whereon to feed....

  Have no sense
  That far hence   [20]
Air is sweet in a blue immense,

  Thus, black, blind,
  You have opined
Nothing of your brightest kind;

Never seen [25]
Miles of green,
Smelt the landscape's sweet serene.

192–

Clement's Inn a building, associated with lawyers, in central London. Hardy worked there for a few weeks in 1870.

[12] **ancestry** ancestors.
[15] **swarthy** made dirty by the smoke of London.
[23] **opined Nothing** formed no opinion.

The Fallow Deer at the Lonely House

ONE without looks in tonight
Through the curtain-chink
From the sheet of glistening white;
One without looks in tonight
As we sit and think [5]
By the fender-brink.

We do not discern those eyes
Watching in the snow;
Lit by lamps of rosy dyes
We do not discern those eyes [10]
Wondering, aglow,
Fourfooted, tiptoe.

Deer still come to the windows of Hardy's cottage at Higher Bockhampton. Hardy's love of animals can be felt even in this slight, but poetic, incident.
Fallow Deer a species of smaller deer of a pale brown or reddish yellow colour.

[6] **By . . . brink** close to the fender around the fire.

The Faithful Swallow

WHEN summer shone
Its sweetest on
An August day,
'Here evermore,'
I said, 'I'll stay; [5]
Not go away
To another shore
As fickle they!'

December came:
'Twas not the same! [10]
I did not know
Fidelity
Would serve me so.
Frost, hunger, snow;
And now, ah me, [15]
Too late to go!

Swallows normally migrate southwards from England in autumn To stay longer is dangerous.

The Robin

WHEN up aloft
I fly and fly,
I see in pools
The shining sky,
And a happy bird [5]
Am I, am I!

When I descend
Towards their brink
I stand, and look,
And stoop, and drink, [10]
And bathe my wings,
And chink and prink.

When winter frost
Makes earth as steel
I search and search [15]
But find no meal,
And most unhappy
Then I feel.

But when it lasts,
And snows still fall, [20]
I get to feel
No grief at all,
For I turn to a cold stiff
Feathery ball!

[12] **chink** Hardy's word for the sound made by robins.
prink trim its feathers with its beak.

The Puzzled Game-birds
(Triolet)

THEY are not those who used to feed us
When we were young – they cannot be –
These shapes that now bereave and bleed us?
They are not those who used to feed us,
For did we then cry, they would heed us. [5]
– If hearts can house such treachery
They are not those who used to feed us
When we were young – they cannot be!

Hardy was horrified by man's cruelty to animals. Here he refers to the practice of breeding birds in the spring in order to shoot them in the autumn as part of what is called sport. Once when out cycling with a friend Hardy saw an injured blackbird. His friend described how, 'Hardy shuddered on seeing it. Then he said, "I cannot bear such a sight, poor thing! Could you put it out of its misery? If left, a cat or stoat will get it and torture it." . . . It was not merely a common humanitarian instinct that had been aroused in him, I could see; but a sheer sickness and revulsion against a helpless thing suffering.'

triolet a poetic form which requires an eight-line stanza with two rhymes. The first line is repeated as the fourth and seventh, and the second and eighth lines are alike.
[3] **bereave and bleed** kill and wound.

The Blinded Bird

So zestfully canst thou sing?
And all this indignity,
With God's consent, on thee!
Blinded ere yet a-wing
By the red-hot needle thou, [5]
I stand and wonder how
So zestfully thou canst sing!

Resenting not such wrong,
Thy grievous pain forgot,
Eternal dark thy lot, [10]
Groping thy whole life long,
After that stab of fire;
Enjailed in pitiless wire;
Resenting not such wrong!

Who hath charity? This bird. [15]
Who suffereth long and is kind,
Is not provoked, though blind
And alive ensepulchred?
Who hopeth, endureth all things?
Who thinketh no evil, but sings? [20]
Who is divine? This bird.

Hardy refers to the cruel practice of blinding birds with a red-hot needle because of the knowledge that this causes them to sing better. He criticises a so-called Christian society which allows this to happen, and he emphasises the point by echoing in his final stanza the well-known words of St Paul: 'Charity suffereth long, and is kind; charity envieth not; charity vaunteth not itself, is not puffed up, doth not behave itself unseemly, seeketh not her own, is not easily provoked, thinketh no evil; rejoiceth not in iniquity, but rejoiceth in the truth; beareth all things, believeth all things, hopeth all things, endureth all things.' (1 Corinthians 13:4–7).

Horses Aboard

HORSES in horsecloths stand in a row
On board the huge ship that at last lets go:
Whither are they sailing? They do not know,
Nor what for, nor how. –
They are horses of war,
And are going to where there is fighting afar; [5]
But they gaze through their eye-holes unwitting they are,
And that in some wilderness, gaunt and ghast,
Their bones will bleach ere a year has passed,
And the item be as 'war-waste' classed. –
And when the band booms, and the folk say 'Good-bye!' [10]
And the shore slides astern, they appear wrenched awry
From the scheme Nature planned for them, – wondering why.

A Sheep Fair

THE day arrives of the autumn fair,
And torrents fall,
Though sheep in throngs are gathered there,
Ten thousand all,
Sodden, with hurdles round them reared: [5]
And, lot by lot, the pens are cleared,
And the auctioneer wrings out his beard,
And wipes his book, bedrenched and smeared,
And rakes the rain from his face with the edge of his hand,
As torrents fall. [10]

The wool of the ewes is like a sponge
With the daylong rain:
Jammed tight, to turn, or lie, or lunge,
They strive in vain.

Their horns are soft as finger-nails, [15]
Their shepherds reek against the rails,
The tied dogs soak with tucked-in tails,
The buyers' hat-brims fill like pails,
Which spill small cascades when they shift their stand
In the daylong rain. [20]

POSTSCRIPT

Time has trailed lengthily since met
At Pummery Fair
Those panting thousands in their wet
And woolly wear:
And every flock long since has bled, [25]
And all the dripping buyers have sped,
And the hoarse auctioneer is dead,
Who 'Going – going!' so often said,
As he consigned to doom each meek, mewed band
At Pummery Fair. [30]

[5] **hurdles** movable fences made up of frames of interlaced wood.
[16] **reek** steam.
[30] **Pummery** Hardy's name for Poundbury, an ancient earthwork immediately northwest of Dorchester.

The Roman Gravemounds

BY Rome's dim relics there walks a man,
Eyes bent; and he carries a basket and spade;
I guess what impels him to scrape and scan;
Yea, his dreams of that Empire long decayed.

'Vast was Rome,' he must muse, 'in the world's regard, [5]
Vast it looms there still, vast it ever will be';
And he stoops as to dig and unmine some shard
Left by those who are held in such memory.

But no; in his basket, see, he has brought
A little white furred thing, stiff of limb, [10]
Whose life never won from the world a thought;
It is this, and not Rome, that is moving him.

And to make it a grave he has come to the spot,
And he delves in the ancient dead's long home;
Their fames, their achievements, the man knows not; [15]
The furred thing is all to him – nothing Rome!

'Here say you that Caesar's warriors lie? –
But my little white cat was my only friend!
Could she but live, might the record die
Of Caesar, his legions, his aims, his end!' [20]

Well, Rome's long rule here is oft and again
A theme for the sages of history,
And the small furred life was worth no one's pen;
Yet its mourner's mood has a charm for me.

November 1910

Hardy's favourite cat died in 1910 and was buried in the garden of Max Gate about the time he wrote this poem. It provides an excellent illustration of his practice of mixing fact with fiction.

[7] **shard** broken piece of pottery.
[22] **sages** wise men.

Last Words to a Dumb Friend

PET was never mourned as you,
Purrer of the spotless hue,
Plumy tail, and wistful gaze
While you humoured our queer ways,
Or outshrilled your morning call [5]
Up the stairs and through the hall –
Foot suspended in its fall –

While, expectant, you would stand
Arched, to meet the stroking hand;
Till your way you chose to wend [10]
Yonder, to your tragic end.

Never another pet for me!
Let your place all vacant be;
Better blankness day by day
Than companion torn away. [15]
Better bid his memory fade,
Better blot each mark he made,
Selfishly escape distress
By contrived forgetfulness,
Than preserve his prints to make [20]
Every morn and eve an ache.

From the chair whereon he sat
Sweep his fur, nor wince thereat;
Rake his little pathways out
Mid the bushes roundabout; [25]
Smooth away his talons' mark
From the claw-worn pine-tree bark,
Where he climbed as dusk embrowned,
Waiting us who loitered round.

Strange it is this speechless thing, [30]
Subject to our mastering,
Subject for his life and food
To our gift, and time, and mood;
Timid pensioner of us Powers,
His existence ruled by ours, [35]
Should – by crossing at a breath
Into safe and shielded death,
By the merely taking hence
Of his insignificance –
Loom as largened to the sense, [40]
Shape as part, above man's will,
Of the Imperturbable.

As a prisoner, flight debarred,
Exercising in a yard,
Still retain I, troubled, shaken, [45]
Mean estate, by him forsaken;
And this home, which scarcely took
Impress from his little look,
By his faring to the Dim
Grows all eloquent of him. [50]

Housemate, I can think you still
Bounding to the window-sill,
Over which I vaguely see
Your small mound beneath the tree,
Showing in the autumn shade [55]
That you moulder where you played.

2 October, 1904

'First there is the selectively detailed materialisation of what it was that died: purrer of the spotless hue with the plumy tail, that would stand arched to meet the stroking hand. After the tenderness of immediate memory comes the first reaction: never to risk it again. Then come eight lines which envisage what must be done, and the impossibility of doing it, to blot the memory out. All this Hardy supplied, as it were, by a series of directly felt observations, and these, in their turn, released one of those deeply honest, creative visions of man in relation to death which summoned the full imagination in Hardy as nothing else could.' (R. P. Blackmur)

Dead 'Wessex' the Dog to the Household

Do you think of me at all,
    Wistful ones?
Do you think of me at all
    As if nigh?
Do you think of me at all [5]
At the creep of evenfall,
Or when the sky-birds call
    As they fly?

Do you look for me at times,
    Wistful ones? [10]
Do you look for me at times
    Strained and still?
Do you look for me at times,
When the hour for walking chimes,
On that grassy path that climbs [15]
    Up the hill?

You may hear a jump or trot,
    Wistful ones,
You may hear a jump or trot –
    Mine, as 'twere – [20]
You may hear a jump or trot
On the stair or path or plot;
But I shall cause it not,
    Be not there.

Should you call as when I knew you, [25]
    Wistful ones,
Should you call as when I knew you,
    Shared your home;
Should you call as when I knew you,
I shall not turn to view you, [30]
I shall not listen to you,
    Shall not come.

Hardy had many pets during this long life. Wessex, a wire-haired terrier, was one of the last. He died at Max Gate and was buried in the garden. Hardy designed a gravestone for him which was engraved, 'The Famous Dog, Wessex, August 1913–27 December 1926. Faithful. Unflinching.'

'Ah, Are You Digging on My Grave?'

'Ah, are you digging on my grave,
    My loved one? – planting rue?'
– 'No: yesterday he went to wed
One of the brightest wealth has bred.
"It cannot hurt her now," he said, [5]
    "That I should not be true." '

'Then who is digging on my grave?
    My nearest dearest kin?'
– 'Ah, no: they sit and think, "What use!
What good will planting flowers produce? [10]
No tendance of her mound can loose
    Her spirit from Death's gin." '

'But some one digs upon my grave?
    My enemy? – prodding sly?'
– 'Nay: when she heard you had passed the Gate [15]
That shuts on all flesh soon or late,
She thought you no more worth her hate,
    And cares not where you lie.'

'Then, who is digging on my grave?
    Say – since I have not guessed!' [20]
– 'O it is I, my mistress dear,
Your little dog, who still lives near,
And much I hope my movements here
    Have not disturbed your rest?'

'Ah, yes! *You* dig upon my grave . . . [25]
    Why flashed it not on me
That one true heart was left behind!
What feeling do we ever find
To equal among human kind
    A dog's fidelity!' [30]

'Mistress, I dug upon your grave
    To bury a bone, in case
I should be hungry near this spot
When passing on my daily trot.
I am sorry, but I quite forgot [35]
    It was your resting-place.'

'There is a certain astringent quality in some poetry which I take to be a valuable corrective to the sentimentality that is latent in most of us. "Ah, Are You Digging on My Grave" is a level, bitter poem without grace of style or diction, a little awkward, like Hardy himself . . . think of the sentiment lavished on pet dogs by their owners; as you know a common symptom of psychological incompleteness.' (T. R. Henn) Much as Hardy loved animals his strong ironic sense saved him from sentimentality.

[2] **rue** an evergreen shrub, traditionally the symbol of sad remembrance.
[11] **tendance** tending.
[12] **gin** trap.

Index of Titles

Index of First Lines

Acknowledgements

The publishers wish to thank the following, who have kindly given permission for the illustrations to be reproduced:

The Trustees of the Dorset County Museum: Plate 1
The Trustees of the Thomas Hardy Memorial Collection: Plates 2, 3, 5–6, 7, 10
Birmingham City Museum and Art Gallery: Plate 9
James Gibson: Plates 4 and 8